T0229030

NETWORKS AND GRAPHS:
Techniques And Computational Methods

ABOUT OUR AUTHOR

David K. Smith has degrees from the Universities of Cambridge and Lancaster (in mathematics and operational research respectively); he learnt computing as a member of a research team in low temperature physics, and later researched in water supply modelling for environmental sciences. As a sabbatical, he spent several months teaching at the University of Jordan, in the Faculty of Economics.

In addition to his two earlier books Dr Smith has published many papers in international journals on aspects of several of his research interests, and has presented his work at conferences in several countries. He is the editor in chief of *International Abstracts in Operations Research*, and has played a leading part in planning the electronic versions of this, the leading abstracting journal serving the operational research and management science communities. He is an editorial advisor to three leading journals for the OR/MS community.

Dr Smith is a lay preacher in the Church of England. His hobbies include philately, leading to research articles about aspects of postal history, and travel, both in the UK and abroad. David Smith is a senior lecturer in the School of Mathematical Sciences at the University of Exeter in the United Kingdom. He specialises in teaching and studying networks and graphs, with their applications, and in the use of dynamic programming as a tool for structuring and solving problems. Dr Smith is also interested in the use of operational research in the not-for-profit sector and in developing countries.

Networks and Graphs:
Techniques and Computational
Methods

David K. Smith
University of Exeter

Horwood Publishing
Chichester

HORWOOD PUBLISHING LIMITED
International Publishers in Science and
Technology, Coll House, Westergate,
Chichester, West Sussex, PO20 3QL, England

COPYRIGHT NOTICE
All Rights Reserved. No part of this publication may be reproduced, stored
in a retrieval system, or transmitted, in any form or by any means,
electronic, mechanical, photocopying, recording, or otherwise, without the
permission of Horwood Publishing Limited.

© David K. Smith, 2003

British Library Cataloguing in Publication Data
A catalogue record of this book is available from the British Library

ISBN 1-898563-91-8

Table of Contents

Preface

In the Old Testament of the Bible, Chapter 8 of the book of *Joshua* records that:

> (3) So Joshua and the whole army moved out to attack Ai. He chose thirty thousand of his best fighting men and sent them out at night (4) with these orders: "Listen carefully. You are to set an ambush behind the city. Don't go very far from it. All of you be on the alert. (5) I and all those with me will advance on the city, and when the men come out against us, as they did before, we will flee from them. (6) They will pursue us until we have lured them away from the city, for they will say, 'They are running away from us as they did before.' So when we flee from them, (7) you are to rise up from ambush and take the city. The LORD your God will give it into your hand.

However, the Bible doesn't record the conversation which followed:

> Joshua drew a plan in the dust to show the city, the place of the ambush and the route to be taken. The leader of the fighting men asked: "How do you know that the route is the shortest? Are you sure that we can move thirty thousand men during the dark? Will there be enough provisions and water for us?" And Joshua said: "These are the instructions that you must follow. If you want a mathematical proof that the route is best, and that you can do it, then you will have to wait three thousand years."

This book is concerned with the three problems in that imaginary conversation. How do you find the shortest (or the best) route between two points? How do you move people (or goods) when there are restrictions on their movement? What resources are consumed by such movements?

The book also looks at many problems that are connected with these. For many years, humans have used diagrams to represent the world around them. Often, the diagram is also used to help answer some question about "what is best?" in some way of measuring "best". This book is concerned with the kind of diagram which mathematicians call a graph or a network. The subject is of interest to mathematicians, because of the beauty of the mathematical area of study known as graph theory; it is of interest to engineers, because many graphs and networks represent some engineering project; it is of interest to operational research scientists, since modelling the real world is an essential part of their vocation; and it is of interest to everyone else, because, as the book demonstrates, everyday problems of transport, communication and finance can often be solved using graphs and networks.

Robert Louis Stevenson wrote: "Every book is, in an intimate sense, a circular letter to the friends of him who writes it". This is therefore a letter to friends across the world, today, tomorrow, and for the future. I hope that you enjoy it.

Now some notes for specific types of readers:

Teachers and lecturers

This book will cover a semester's teaching programme concerned with the theory and application of graphs and networks. The ordering of the chapters is slightly arbitrary: it would be possible to present the material in another sequence without serious problems. The order that I have chosen is one which has proved reasonably logical over the years that I have been teaching it. There is a selection of exercises which deal with understanding basic concepts of the theory and definitions, and also concerned with the process of selecting an appropriate model to use. In places, the exercises bring together material from different chapters, to illustrate and develop the idea that this is an area of study which is integrated. The worked examples in the text have been chosen to be small enough to deal with in print without being boring and yet big enough that they are not trivial. As a teacher, you may find it worthwhile developing your own simple examples of (about) the same size, to give variety for your classes. Hints for selected exercises are at the back of the book.

Students

Thank you for buying or borrowing this book. If you have borrowed a copy, I hope that one day you will feel able to recommend that it should be purchased by your future employer as an essential reference guide. I also hope that you find the book readable and comprehensive. Much of the material is based on courses that have been taught at the University of Exeter in the United Kingdom, and students have given feedback about what is simple to understand and what is not.

Reviewers

You may be in one of the categories above as well, but your job is to write a few hundred words for a learned journal (textbooks like this are never reviewed for the mass media) about this book. Your reward is to keep the book—so I hope that it is of sufficient interest that it doesn't collect dust on your shelves, nor that it is sold to a second-hand book dealer. Please feel free to use such phrases as: "This is a fascinating book."; "The author is enthusiastic."; "It is not so comprehensive as ... (insert your favourite 1000-page text-book)". If you don't have time to read the whole book, then:

- You should have taken it on a longer train journey;

- You shouldn't have fallen asleep in the sun in your garden chair;

- You might use Donald Knuth's sampling scheme, and read the whole of every tenth or twentieth page, and judge the book on that basis.

Acknowledgements

It is customary for authors to express their thanks to those who have helped them in any way. My colleagues and students have encouraged me as a teacher and writer: it would be invidious to single out any of them by name, but my gratitude is there nonetheless. This book was written using TeX and LaTeX, and I am grateful to those who developed these document processing and type-setting systems. Many of the diagrams were produced using Richard Nickalls' filter program, MathsPIC.

Finally, in this list of thanks, for her steadfast support, my wife Tina's name should be here in letters of gold.

1

Introduction

When you have read this chapter, you should be able to:

- explain what is meant when mathematicians and operational research (O.R.) scientists refer to a graph or network;

- explain the terms: *graph, network, edge, vertex, arc, node, weight, path* and *algorithm*;

- recognize the variety of real-life problems which can be studied using graphs and networks;

- appreciate that solving some such problems may take a very long time.

1.1 GRAPHS AND NETWORKS

Graphs, in their mathematical sense, are a special case of a wider concept called **multigraphs**. A multigraph is formed from two parts. First, there is a set of points, called **nodes** or **vertices**, and second there is a set of lines which join pairs of these points. These lines are known as **arcs** or **edges**. For convenience, this combination will be referred to as G, or $G = (V, E)$ with a **vertex set** V and an **edge set** E with $|V|$ vertices and $|E|$ edges. Graphs are multigraphs in which no two edges "connect" the same vertices, and in which each edge connects different vertices. Within the covers of this book, multigraphs are of very little importance, and the focus will almost invariably be on graphs.

Often, a graph is drawn with its edges as straight lines, but there is no need for this. A diagram of a graph is simply one way of showing its **topology**. An edge can be described by identifying the vertices that it joins; in many graphs, the edges will be directed from one vertex to the other, giving a directed edge. When all the edges in a graph are directed, then the resulting **directed graph** is referred to as a **digraph**. When there are no directed edges, then one has an **undirected graph** and the third possibility—some edges directed and the remainder undirected—is a **mixed graph**. In this book, the sets of vertices and edges are finite ones; some research work is devoted to the study of infinite graphs and their properties.

Graphs are used to represent many everyday phenomena. Most railway and underground systems around the world use a graph to show the lines and the stations, with emphasis on the places where it is possible to change between train services. The vertices are these places, the edges, the lines. Diagrams of roads, streets, canals and airline routes are similar. Gas, water and sewage pipe systems, and electricity distribution cables (both inside a building and between power station and consumer) are often shown by graphs with vertices representing places where two or more pipes and cables meet. Recruitment brochures for many companies show the "promotion ladder" using a graph, with vertices for the jobs and edges showing how employees may progress from one to another. Within companies and organisations, graphs are used to show areas of responsibility, with a directed edge showing which people and committees report to others. Trees, a particular form of graph, are used to show the hierarchy of files and directories on a computer, and to trace a family tree.

Graphs which represent railway networks are generally undirected; most of the time, if one can travel from one station to another by rail, then one can make the return journey. But graphs representing roads and streets in a town centre will often be mixed, because there will be one-way streets. Graphs of water pipelines will generally be directed, since the water can only flow in one direction in the pipelines which are represented by the edges.

A **network** is a graph on which one has defined additional information, usually numerical data, associated with the edges (and sometimes with the vertices). The data may be (as the most common examples) distances, costs, times, limits on the flow of material through the edge or vertex, or a combination of these properties. The term **weighted graph** is used when there is only one piece of numerical data on each edge, and that data can be added together to give a value for a collection of edges, such as the cost or the length of the edges.

In general, the size of a network (and the graph it is based on) will be measured by the number of vertices, often shown as n. This is a convenient shorthand for the mathematical formulation of a graph, $G = (V, E)$, and the number of components $|V|$ and $|E|$. In this book, most of the examples will use small graphs and networks; in reality, networks of practical use will have hundreds or thousands of edges and vertices.

1.2 ALGORITHMS

Many of the examples of networks will be so small that any problems can be solved by trial and error, using pencil and paper and the diagram of the graph. However, when the network is large enough to be of interest, then these simple methods will not work. Then one needs precise and systematic rules, which will be easy to explain to other people, and which give some kind of promise that the answer is what one wanted in the first place. A step-by-step method for solving any problem is an **algorithm**. In this book, many algorithms will be written in a way that could easily be converted to computer instructions; this is deliberate, because generally computers are the tools needed for solving large graph or network problems.

There are everyday calculations that use algorithms, although one may not always realize that one is using a set of systematic rules. For instance, summing a set of numbers, say 13, 74, 83, 49, on a calculator, one follows the rules below:

0: (Initialisation step, done once only.) Enter the first number in the display;

1: If there are no more numbers, then type "=" and stop with the answer; otherwise type "+" and the next number;

2: Go back to step 1.

However, this is probably not the same algorithm which one learnt as a small child, doing such simple calculations on paper. Then the process was more complex to describe, because one separated the numbers into their "tens" and their "units", added the "units" together to give the last digit of the answer, carried some "tens" and added the "tens" together. Both methods are acceptable algorithms for summing a set of numbers, and the choice is a reminder that there may be several ways of obtaining the same result in a problem.

Whenever one uses or studies an algorithm, it should have five attributes:

1. **Finiteness:** It must stop after a finite number of steps, and so should have a rule which will make it terminate under circumstances which the writer or user knows will happen sooner or later.

2. **Definiteness:** Each step must be defined precisely. The user (or the user's computer) must have clear instructions on what to do in any circumstance. Algorithms are presented in English, but it should be understood that they will be translated into a computer language for everyday use.

3. **Input:** There must be quantities that are given to it beforehand; usually these will be those that the algorithm works on, but there are a few cases where there is no input. The input for many problems will be the graph and the parameters of the network. (Later in the book, there is a discussion about how data can be represented in computers for graph and network problems.)

4. **Output:** All algorithms have one or more outputs, which may be one or more numbers, or some information. The information may be that the algorithm has stopped and the reason why. The rule about finiteness means that the algorithm will always stop, and often it is important to know why this has happened.

5. **Effectiveness:** In theory, one should be able to follow all the steps of an algorithm using pencil and paper; sometimes the worked examples will do that later in this book, and these examples may seem boring. However, if one actually works through the examples with the algorithms, then it will be easier to see what is happening.

1.3   BASIC DEFINITIONS

The essential idea of a graph is an extremely simple one. As soon as one starts to examine them, one finds that graphs have numerous properties and these lead to a large number of definitions. Suppose that a graph has an edge $e = \{p, q\}$. Then p and q are the **ends** or **end vertex** of the arc. The vertices p and q are **incident with** the edge e, and these are **neighbours** of each other, or that these are **adjacent** vertices. The edge e is **incident to** p and q. The number of edges which are incident to a vertex p defines the **degree** of vertex v, often written $d(p)$. An edge which is a loop (p, p) counts twice for the calculation of the degree of p. (For the most part, these names and other terms in the book have obvious meanings.)

There are many sorts of graph which are so important that they merit their own names. Here we consider two, relating to the edges which connect pairs of vertices. First, a **complete graph** is one which has every possible edge. If v_i and v_j are two vertices in a complete graph, then there will be an edge (v_i, v_j). So, if there are n vertices, there will be $n(n-1)/2$ edges in the complete graph. This graph is often given the symbol K_n. Second, a **bipartite graph** is one in which the set of vertices can be divided into two, V_1 and V_2; the edges (v_i, v_j) all satisfy $v_i \in V_1, v_j \in V_2$, so there are no vertices in V_1 or V_2 which are neighbours of each other. Such a graph is a **complete bipartite graph** if all the possible edges are present. Often, complete bipartite graphs with m vertices in one set and $n - m$ in the other will have the symbol $K_{m,n-m}$. Some examples are shown in Figure 1.1. A notation which is sometimes used for bipartite graphs is to describe them as $G = (V_1, V_2, E)$.

In a diagram of a graph like this, the edges may cross one another. Points where this happens are not significant; they are not vertices of the graph, and are simply a result of trying to give a clear visual picture of the complexities of the graph. It is generally advisable to draw diagrams where only two edges cross at a point. Drawing tidy graphs given the sets of vertices and edges is an art, and one often needs two or three attempts to produce a clear and neat picture.

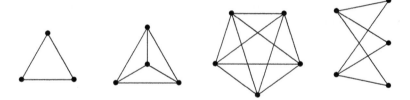

Figure 1.1: The complete graphs, K_3, K_4, K_5, and the complete bipartite graph $K_{2,3}$

1.3.1   Paths, cycles and connectedness

It is usual to represent the members of the vertex set V by positive integers, $1, 2, 3, \ldots, n$, which means that they can be numbered and counted in the everyday sense. A typical vertex will be called v_j. Similarly, a typical edge will be e_i, indicating an ordered pair of distinct vertices (v_j, v_k). The first properties of graphs are concerned with what happens when several edges are linked together.

Suppose one takes a sequence $e_1, e_2, \ldots, e_r$ of edges, which have the property that the one end of edge e_i is also an end of e_{i+1}. Such a set will define a sequence of vertices, $v_0, v_1, \ldots, v_r$, with $e_i = (v_{i-1}, v_i)$ (or (v_i, v_{i-1})). These equivalent sequences define a **walk** in a multigraph, and this may also be defined as the sequence $v_0, e_1, v_1, \ldots, e_r, v_r$. Walks which have all vertices distinct are **paths**, with a **starting** or **initial** vertex v_0 and **final** or **terminal** vertex v_r. If all the edges are distinct, then one has a **trail**. A path in a multigraph whose start and finish are the same is known as a **circuit**. In a graph it is known as a **cycle**. In a graph, a path has **length** r units, since this is the number of edges. In a few situations, all that matters is whether or not the length is **odd** or **even**. The **distance** between two vertices, v_1, v_2, written $d(v_1, v_2)$ is the smallest number of edges in a walk between these vertices. The **diameter** of a graph is the largest distance between any pair of vertices. The **length** of a cycle, and its oddness or evenness, follows the same basic idea as for a path.

This definition of a path applies to all kinds of graph. When the path is made up of edges which all run in the same direction, from the initial vertex to the terminal vertex, then the result is a **directed path**. A path in a directed graph is also sometimes referred to as a **chain**.

A graph may be **connected** or (guess what!) **disconnected**. Connected graphs are those that possess a path between every pair of vertices. Disconnected graphs do not. One way of describing a disconnected graphs is to view them as being made up of a set of separate **components**. Removing an edge from a graph will either leave the number of components unchanged, or increase it by one. An edge whose deletion increases the number of components by 1 is a **cut-edge**. Figure 1.2 has two cut-edges, (2,4) and (5,7).

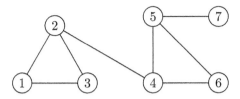

Figure 1.2: A graph with two potential cut-edges, (2,4) and (5,7).

An extension of this concept of cut-edges permits one to define a **cut** in the graph as follows. A cut is a set of edges whose removal increases the number of components of the graph by 1. Sometimes this definition is made even more specific by stating that the cut **separates** two specified vertices, i and j, which had been in the same component and are now in different ones. For instance, in Figure 1.2, the set of edges $\{(4,5),(4,6)\}$ is a cut which separates vertices 1 and 7. A cut which converts a connected graph into two components is the most frequently encountered type.

1.3.2   Subgraphs and trees

Suppose that a graph $G = (V, E)$ has been defined by its sets of vertices, V, and edges, E. What happens when some of the members of one or both of these sets are removed? If one takes a subset, V_1 of V, then some of the edges in E cannot exist any longer. Take $E|V_1$ to mean those members of E which still exist, because both their ends are in V_1. Then the sets $V_1, E|V_1$ define a new graph, which is known as the **induced subgraph** on V_1. This is often denoted $G|V_1$. Taking a subset of the edges in an induced subgraph, with a set $E_1 \subset E|V_1$, creates a **subgraph**. A special case occurs when one takes the original set of vertices and a subset of the edges, giving a graph (V, E_1) with $E_1 \subset E$. This is a **spanning subgraph**.

Induced subgraphs have links to the concept of cutting vertices. A **vertex-cut** in a multigraph $G = (V, E)$ is a set of vertices U whose removal creates an induced subgraph with more components than G. If the multigraph represents some form of communication system, a vertex-cut would prevent communication between all vertices. If the set U consists of a single vertex, then that is known as a **cut-vertex**. Vertices 2, 4 and 5 in Figure 1.2 are possible cut-vertices.

A graph without any cycles is known as a **forest**. If the graph is connected then one has a **tree**. Spanning subgraphs which are also trees are of considerable importance in the applications of graphs and networks, and they will be discussed in detail later on.

In Figure 1.3 there is a graph with seven vertices and twelve edges; the vertices represent places in Britain's "West Country". This graph could be defined by the vertex set $V = \{$Arne View, Bindon, Collepardo, Drizzlecombe, Exeter, Fairfield, Glencot$\}$, abbreviated to $V = \{$A,B,C,D,Ex,F,G$\}$ with edge set: $E = \{$(A,B), (B,C), (C,D), (D,F), (F,G), (G,A), (A,Ex), (B,Ex), (C,Ex), (D,Ex), (F,Ex), (G,Ex)$\}$. Alongside each edge, using the notation that is common, there is the numerical value of the property associated with the edge in the network; in this case it is the road distance measured in kilometres.

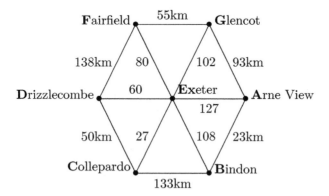

Figure 1.3: A network representing 7 locations in England's West Country, with the road distances between them.

1.4   COMPLEXITY OF ALGORITHMS

In some of the analysis that follows, there will be reference to the **computational complexity of algorithms.** This is an idea which has been developed in computational mathematics, to try and compare the efficiency of different methods for solving the same problem. Suppose that one has two networks, one with n nodes, the other with $2n$. Suppose that there is an algorithm which performs some task on the network. It would be reasonable to expect that the algorithm will take longer, on average, to perform that task on the larger network. But how much longer?

The answer is usually given in terms of the notation $O()$. An algorithm takes a time of order $t(n)$ (written as $O(t(n))$) if there is a function $t()$ and a positive constant c (independent of n) such that the algorithm can solve the problem in a time less than $ct(n)$ for all values of n.

For example, if one is given a list of n numbers and asked to find the smallest, then the time taken will be $O(n)$ because all that one needs to do is to keep a note of the smallest that has been found so far, and compare it with all the numbers that appear later in the list. But if one wanted to put the numbers into ascending order, then more work must be done; the (naive) method of finding the smallest, then the next smallest and so on will require $n + (n-1) + (n-2) + (n-3) \ldots + 2 + 1 = n(n+1)/2$ operations. This analysis provides a function $t(n) = 0.5n^2 + 0.5n$, with $c = 1$; it is much more convenient to say that $t(n) = n^2$ and $c = 1$, since $n^2 \geq 0.5n^2 + 0.5n$ for all positive integers. (Alternatively, $O(n^2)$ is **dominant** over $O(n)$.) Hence this method of sorting is $O(n^2)$. (Can one sort numbers more quickly?) The algorithm given earlier for adding n numbers in a calculator has complexity $O(n)$; the rule for adding integers by adding the "units", and then the "tens", "hundreds" and so on will take longer, and its complexity is $O(n \log_{10}(M))$, where M is the largest integer in the set.

Most algorithms that are encountered in operational research have computational complexity which is $O(n^r)$ for $r = 1, 2, 3, \ldots$. (As has been shown in the last paragraph, if an algorithm has a complexity which is the sum of several terms, then one should express it in terms of the dominant function.) The simplex algorithm for linear programming doesn't. It has **exponential complexity** which means that the time for finding a solution cannot be bounded except by a function such as 2^n.

Analysts classify problems as P, NP and NP-complete, and describe some as being NP-hard. Those referred to as P are those for which there is an algorithm whose complexity is polynomial. The NP-problems are those for which—as yet—there is no known polynomial algorithm. The NP-complete problems are in a subset of the NP problems; any NP problem can be changed into an NP-complete one using an algorithm whose complexity is polynomial. So if one could find a polynomial algorithm for any NP-complete problem, then all NP problems could be solved in polynomial time. One of the unsolved questions facing mathematics in the twenty-first century is: "Is the set P equal to the set NP?". (There are many books and papers relating to computational complexity in general and this question in particular. One which deals with algorithms similar to those in this book is that by Gregory Rawlins ([18]).)

1.5   OPTIMISATION

Graphs and their properties offer considerable scope for mathematical research. This book is concerned with finding the best (or optimal) properties of graphs, where "best" will be defined in a suitable way in each case. This will generally mean defining some objective function whose value is to be maximised (made as large as possible) or minimised (as small as possible) by changing some variable(s) which are network properties. Maximisation and minimisation are effectively the same problem, because making a function $f(x)$ as large as possible is equivalent to making its negative, $-f(x)$, as small as possible. Optimisation usually requires following a series of steps, possibly many times, and gradually improving the best value that has been found for the objective function. Sometimes these steps may be interrupted and the best value that has been found noted as being acceptably close to the optimum, provided that it is feasible, satisfying any constraints for the problem.

1.6   HEURISTICS.

There will be several passing reference to heuristics in this book. The term will be used to mean a set of rules which aim to find a good solution to the problem being studied, but which have no guarantee of optimality.

> "... a procedure for solving problems by an intuitive approach in which the structure of the problem can be interpreted and exploited intelligently to obtain a reasonable solution."

This means that a heuristic method will take a problem, and find a **feasible** solution to it. This solution will satisfy the constraints of the problem, but it may not be the best possible solution. (In everyday life, heuristics provide the rules of thumb which people use to solve both simple and complex problems without needing to perform a huge amount of analysis. Question: "Which queue do you join in the supermarket?"; Answer: "The shortest, measured in terms of people or items in their baskets." Question: "How many ball-point pens do you take to a three-hour university examination?"; Answer: "At least one more than you are ever likely to need, allowing for them to run out of ink, or to prove defective.")

1.7   INTEGER PROGRAMMES

In the same way, there are a few references in this book to integer programmes. It is not necessary for readers to know how to solve integer programming problems. The only use that is made of linear and integer programmes is in the formulation of some problems with numerical variables. For convenience, key features of integer programmes are:

> An **integer programme** is like a linear programme, that is it has:
>
> - One objective function which is a linear function of the decision variables to be optimised (either maximised or minimised);
> - At least one constraint; all constraints are linear functions of the decision variables;
> - Decision variables which are non-negative.
>
> **and**   at least one decision variable is required to take an integer value, not a real value.

Variables which are constrained to be either 0 or 1 are sometimes referred to as **binary variables**. These may be used to indicate whether or not some event has happened, so may be called **indicator variables**.

1.8   EXERCISES

1. Find as many examples of graphs or multigraphs which represent something in the real world as possible from printed and electronic sources. Which of these are multigraphs, which are directed, which are mixed? What measures (if any) are associated with the edges?

2. The quizmaster on a TV show has a card with an integer M written on it. To win the prize, you must guess the number as quickly as possible, by guessing a number m, and being told if m is less than M, equal to M, or greater than M. You know that $1 \leq M \leq 1000000$. What is the best strategy for choosing successive values of m?

3. Describe an algorithm whose input is an integer M, followed by a set of M positive integers, $y_1, \ldots, y_M$. The output should be either the difference between the largest and the smallest integers in the set, or a message to say that the data is in error if $M \leq 1$. What is the complexity of your algorithm?

4. The graph $G = (V, E)$ is connected; what is the longest path that is possible on G? And what is the longest trail?

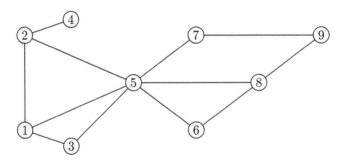

Figure 1.4: A graph with cut-vertices and cut-edges.

5. In the graph of Figure 1.4, find all the cut-vertices, and find cuts with 1, 2 or 3 edges which separate vertex 4 from vertex 9.

2

Trees

When you have read this chapter, you should be able to:

- Recognize when a sub-graph is a tree;

- Be able to construct certain types of trees;

- Understand the applications of trees.

2.1 INTRODUCTION

Among the definitions in the last chapter, the idea of a tree was given. There a tree was defined as a connected graphs with no cycles. This definition implies other properties, examined briefly below. In general, connected graphs will possess many subgraphs which are trees. These are called **trees on the graph**. Study of these is of especial interest, since one may select a tree to possess some characteristic which optimises a property of the tree.

Much of this chapter is concerned with minimal spanning trees, with a section devoted to directed and rooted trees and their uses.

2.1.1 Properties of trees

Starting with the definition of a tree, various properties follow, which give alternative definitions of trees, or ways of recognizing when a graph is a tree. These include:

- A tree T is a connected graph with n vertices and $n - 1$ edges, that is a connected graph with no cycles has one fewer edge than the number of vertices. Assume it is true that a tree with $n - 1$ vertices has $n - 2$ edges; then when one vertex is added to such a tree, one edge must be added to connect this vertex to another. If two edges are added, then there will be a cycle. When $n = 1$, the result is obviously true, so induction gives the general result.

- A tree is a graph with no cycles and $n - 1$ edges. This can be proved inductively in much the same way as before, and so one has the result that any two of the three properties ($n - 1$ edges, no cycles, connected) imply the third.

- T is connected and all its edges are cut-edges. Removing one edge from T will give a graph with one edge too few. If the edge (i, j) is removed, then the vertices can be divided into two sets, those connected to i and those connected to j. The sets are disjoint because there are no cycles, so the arbitrary edge is also a cut-edge. Since (i, j) was a typical edge, every edge is a cut-edge.

- There is one and only one path between any pair of vertices in T. This follows from the definition—there must be a path, since T is connected, and that path is unique because there are no cycles.

- Adding an edge to T will create a graph with one cycle. If the edge (j, k) is added, then the existing path from k to j can be extended back to k to make a cycle—and again this is true for any possible edge (j, k).

Given an undirected graph $G = (V, E)$, a **spanning tree** T on G is defined as a subgraph of G which is also a tree.

2.2   MINIMAL SPANNING TREES

Given $G = (V, E)$, and a weight (a real—or often integer—number, usually positive or zero) $w(e)$ for each edge $e \in E$, the **weight of a spanning tree** T is defined as the sum of the weights of the edges in the tree. (For an edge $e = (i, j)$, the weight $w(e)$ can be written as w_{ij}.) The **minimum-weight spanning tree** or, commonly, the **minimum spanning tree**, is a tree whose total weight is least, taken over all possible spanning trees on the graph.

Finding such a tree is the most commonly encountered problem, but it is not the only type which might be of interest. Recent study of communications networks has highlighted problems where the objective is to find a tree whose maximum degree is less than some specified amount. This is because vertices represent places where several communications channels meet, and practical needs mean that the number of channels coming together at one place must not be too large. Other research work has been dedicated to problems where the number of edges between vertices must be less than some upper bound. But, to start with, in this chapter, minimum spanning trees will be discussed.

The number of spanning trees on a typical graph is extremely large. Cayley ([3]) in 1889, showed that for a complete graph on n vertices, there are n^{n-2} spanning trees (so K_5 has 125 spanning trees, K_6 has 1296, and so on). For other graphs, the number is less, but it does not make sense to consider exhaustively studying each spanning tree to try to find the best.

2.2.1   Applications

Assume that the weights on the edges represent the cost of connecting the corresponding vertices. This may be the cost of building a road, or laying a pipeline. The minimum spanning tree is the cheapest way of ensuring that each vertex is connected to each of the others by a path. Since the tree has no cycles, there will only be one path between any pair of vertices.

In some statistical tests, n observations are made of data from two (or more) sources, giving data $x_1, \ldots, x_n$ and $y_1, \ldots, y_n$. These observations may be thought of as the vertices of two graphs, and the weight of the edges is some measure of the similarity of the two observations at the ends of the edge. The weight of the minimal spanning tree is a measure of the similarity of the data, and so the different sources can be compared.

Minimal spanning trees are also used in other algorithms for graphs and networks, such as the travelling salesperson problem (Chapter 11).

2.2.2   Graph properties of minimal spanning trees

Suppose that a tree T has been defined on a graph $G = (V, E)$.

Take any two vertices, i and j, with the edge $e = (i, j)$ in G but not in T. There is a unique path in T between i and j, and this will combine with the edge e to form a cycle. This cycle is called the **fundamental cycle** of G relative to T with respect to e. It depends on both e and T.

Removing an edge $e = (i, j)$ that is in the tree will separate the vertices into two sets, one containing i and one containing j. There will then be no path between a pair of vertices one in each set (otherwise there would be a cycle in the tree T). The two sets define a **cut** and a **cutset** in G, which is called the **fundamental cutset** of T with respect to the edge s. Once again, this cutset depends on e and T.

Examples of these are shown in Figure 2.1

2.2.3   Optimality conditions for a minimal spanning tree

These definitions are used in the following theorems.

Theorem 1 *A spanning tree T in a weighted graph is a minimum spanning tree if and only if every edge in the tree is a minimum-weight edge in the fundamental cutset defined by that edge.*

Theorem 2 *A spanning tree T in a weighted graph is a minimum spanning tree if and only if every edge not in the tree is a maximum-weight edge in the fundamental cycle defined by that edge.*

2.2.4   Algorithms for a minimal spanning tree

The three best-known algorithms for finding a minimal spanning tree all rely on these two theorems. Each constructs a tree, one step at a time, by selecting

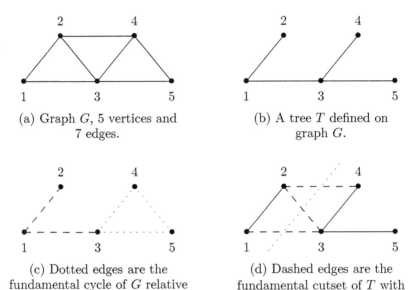

(a) Graph G, 5 vertices and 7 edges.

(b) A tree T defined on graph G.

(c) Dotted edges are the fundamental cycle of G relative to T with respect to (4,5)

(d) Dashed edges are the fundamental cutset of T with respect to edge (1,3)

Figure 2.1:

an edge, including it if it is a suitable small edges, and rejecting it if it is (in some sense) an unsuitable large one.

An attractive way of thinking of this is in terms of **colouring** the edges of the graph. To start with, each edge is uncoloured. Then a step in the construction is to select an uncoloured edge and colour it blue (included) or red (excluded). The decision as to which colour is which follows the rules:

> *Blue rule:* Select a cutset which has no blue edges in it. From the edges which are uncoloured, choose the edge with the smallest weight and make it blue.

> *Red rule:* Select a cycle which has no red edges in it. From the edges which are uncoloured, choose the edge with the largest weight and make it red.

These rules can be applied in any order. The tree will be made up of the blue edges, and during construction of the tree, vertices can be coloured blue or uncoloured, depending on whether or not there is a blue edge incident on them.

2.2.5   Kruskal's algorithm

0: Given the graph $G = (V, E)$; initialize $T = (V, E_T = \emptyset)$. Create a list L of edges from E in ascending order of weight; if there are ties, then rank those edges arbitrarily.

1: Select the edge (i, j) at the head of L and remove it from L. If (i, j) forms a circuit in T, then discard this (colour it red); otherwise add (i, j) to E_T (colour it blue).

2: If T is a tree then stop; otherwise repeat Step 1. (The test for T being a tree can be carried out by calculating $|E_T|$; T will be a tree when $|E_T| = |V| - 1$.)

(In Step 1, the coloured edges mean that it is very easy to decide whether or not there is a circuit. If the uncoloured edge being considered connects two blue vertices from the same sub- tree, then colour it red, otherwise colour it blue.)

Kruskal's method ([14]) works by selecting the edges of least weight, and gradually joining them together into sub-trees which are then assembled to form a tree.

Its complexity is $O(|E| \log(|V|))$, for edges which need to be sorted.

2.2.6   Prim's algorithm

Prim's method ([17]) doesn't use any red edges at all. It selects a vertex and grows one tree from that, one edge at a time.

0: Given the graph $G = (V, E)$ with $n = |V|$; initialize $T = (V, E_T = \emptyset)$. Select any vertex s and colour it blue.

1: Examine all the edges (i, j) for which i is coloured and j is not. Let (k, l) be the edge with smallest weight. Colour this edge blue (which automatically colours the uncoloured vertex as well) and add it to T.

2: If there are $n - 1$ coloured edges (or n coloured vertices), then stop, otherwise repeat Step 1.

The complexity of this method depends on the way that the data relating to weights is structured, but it is possible to achieve the same complexity $(O(|E| \log(|V|)))$ as for Kruskal's method.

2.2.7   Boruvka's algorithm

Kruskal's method can generate sub-trees irregularly across the graph, and these are brought together in an irregular way. Prim's method gradually grows a tree from one vertex, so that many edges are ignored in the early stages of the method. Boruvka's method ([2]) uses edges from the whole graph all the

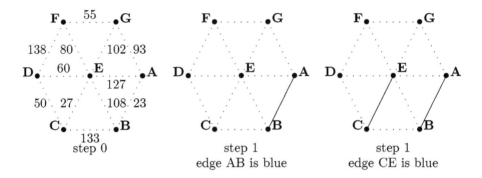

Figure 2.2: Kruskal's method: Step 0 shows the original network, with all edges uncoloured (dotted). Then Step 1 is repeated, with edges being coloured (blue is solid, red is dashed).

time, and extends the sub-trees more uniformly than in Kruskal's approach. (Boruvka's method is also sometimes referred to as Sollin's algorithm, since Sollin ([20]) appears to have rediscovered it independently.)

0: Create a set of sub-trees by colouring each vertex blue.

1: For each sub-tree, select the minimum weight edge that is incident on it. Colour this edge blue.

2: If there is more than one sub-tree left, repeat step 1.

Step 1 of this method reduces the number of separate sub-trees and increases the number of blue edges by at least one, and possibly many more, each time it is performed. Since all the vertices are in one of the sub-trees all the time, this method will make the tree grow more or less uniformly across the whole graph. It too has complexity $O(|E| \log(|V|))$.

2.2.8 The algorithms in practice

Taking the network of places in England's West Country, the three algorithms will (obviously) produce the same minimal spanning tree, because the edge weights are all different, but it will be constructed in different ways. The diagrams which follow show how this happens. Kruskal's algorithm is shown in Figures 2.2, 2.3 and 2.4.

For Prim's method, one needs to specify a starting vertex. In the Figures 2.5, 2.6 and 2.7, vertex G has been chosen.

For Boruvka's method, it is helpful to decide the order in which the vertices will be considered. In this example, alphabetical order is used, as shown in Figures 2.8 and 2.9.

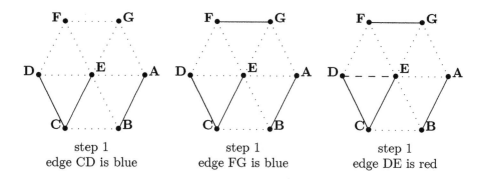

| step 1 | step 1 | step 1 |
| edge CD is blue | edge FG is blue | edge DE is red |

Figure 2.3: When Step 1 is applied for the 5th time, an edge is encountered which makes a circuit, so this edge is coloured red.

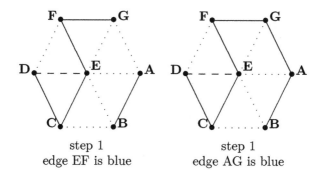

| step 1 | step 1 |
| edge EF is blue | edge AG is blue |

Figure 2.4: Completing the tree with Kruskal's algorithm.

2.2.9   Further aspects of the minimal spanning trees

Uniqueness. Minimal spanning trees are not unique. In each of these algorithms, there are possible choices where edge lengths are equal. For Kruskal's method, the ordering of lengths provides the means for a choice. For Prim's method, the selection of the first vertex was arbitrary, and then the selection of the next edge to colour could give scope for alternative trees being found. Similarly, in Boruvka's method, step 1 selects one edge to colour, which might be of equal length with one or more others. The order in which the vertices are examined also affects the way that the tree grows and possibly which tree will be found if two or more have the same minimal length.

Maximal spanning trees. It is not necessary to devise new methods to find the spanning tree whose total weight is greatest. Maximisation and minimisation of objective functions are equivalent; any of the algorithms can be used,

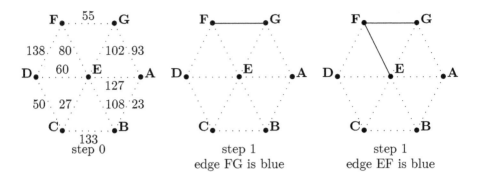

Figure 2.5: Prim's algorithm: The same notation is used as before (uncoloured edges are dotted, blue edges are solid). There are no red edges. G is the initial vertex.

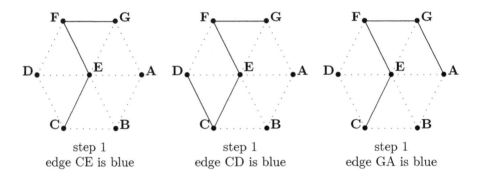

Figure 2.6: Prim's algorithm continuing.

in one of two ways. First, one can replace any reference to smallest or least with one to greatest. Second, one can multiply all the weights by -1, and find the minimal spanning tree in the resulting network.

Other objectives. The minimal spanning tree is also the tree whose maximum edge is as short as possible. This **minimax objective** is sometimes useful for vehicle routing, where one wants to plan routes between vertices so that trucks going between vertices never travel excessive distances. (Applications could be for security, making sure that vehicles can move from one secure depot to another in a short time.) It is possible to prove this result by assuming that edge (k, l) is the longest in a minimal spanning tree and showing that none of the other minimal spanning trees has a shorter edge. The

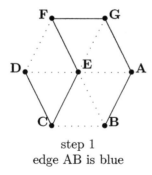

step 1
edge AB is blue

Figure 2.7: Completing the tree with Prim's algorithm.

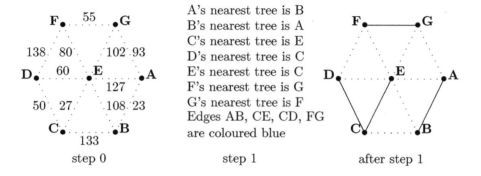

A's nearest tree is B
B's nearest tree is A
C's nearest tree is E
D's nearest tree is C
E's nearest tree is C
F's nearest tree is G
G's nearest tree is F
Edges AB, CE, CD, FG
are coloured blue

step 0   step 1   after step 1

Figure 2.8: Boruvka's algorithm: The first iteration makes four edges blue, and there are three sub-trees ($\{A, B\}, \{C, D, E\}, \{F, G\}$).

same methods can also be used whenever the objective can be transformed into the sum of weights on the edges; but none of the methods is well suited to problems where the objective function can only be calculated once the tree has been created. Networks of pipelines are such an example, since the cost depends on the flow in the edges, and their size, together with the design layout of the vertices. For such problems, there is ongoing research work, often using heuristic methods to explore the feasible trees.

2.2.10 Sensitivity analysis

In many operational research models, there is often some uncertainty about the values of parameters. So it is important to analyze how the answer from an algorithm will change if parameter values change; in some cases, the answer will not change at all, but in others, a small change in one parameter value

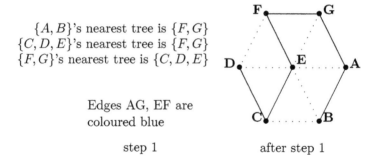

{A, B}'s nearest tree is {F, G}
{C, D, E}'s nearest tree is {F, G}
{F, G}'s nearest tree is {C, D, E}

Edges AG, EF are
coloured blue

step 1   after step 1

Figure 2.9: Boruvka's algorithm: The second iteration makes two more edges blue and completes the tree.

may make a great difference to the solution. This process, sensitivity analysis should be an essential part of the study of any algorithm. However, generally sensitivity analysis is difficult for problems where the answer is a set of discrete objects, such as the output from a minimum spanning tree algorithm.

Assume that the minimal spanning tree is a set of edges, T, of length L_T and that edge (i, j) has its weight w_{ij} changed. As a result, the tree T may be affected; L_T may change as well. There are four cases considered in Table 2.1. The weight of the edge can be increased or decreased, and the edge may or may not be in T.

	$(i, j) \in T$	$(i, j) \notin T$
Increase w_{ij}	(1) First L_T increases, and T is not changed. Then a new tree (possibly) becomes optimal, not using (i, j). If so, (2) applies.	(2) Nothing happens.
Decrease w_{ij}	(3) T_L decreases, T is not changed.	(4) No change to T, until (perhaps) w_{ij} becomes small enough for (i, j) to be in the minimal spanning tree. Then (3).

Table 2.1: Sensitivity analysis for the weight of the minimal spanning tree, and the edges in it, as the length of edge (i, j) is changed.

2.3   ROOTED TREES

For minimal spanning trees, it was assumed that the underlying graph was not directed. In the case of directed graphs, the applications of trees are generally concerned with their practical use rather than the optimality of some property. A **directed tree** is a directed graph whose underlying graph is a tree, in the sense that if the edge directions were to be removed, then a tree would be the result. The most common form of directed tree is a **rooted tree** which is one with an identified vertex (r say) known as the **root** or **root vertex**. There is a unique directed path from r to every other vertex in the rooted tree. Generally rooted trees are drawn with the root at the top, with the direction of edges "down" the diagram, or with the root at the left and edges running towards the right hand side of the page, as shown in the diagrams in Figure 2.10. Arrows are generally not needed in rooted trees, since it should be clear which is the root.

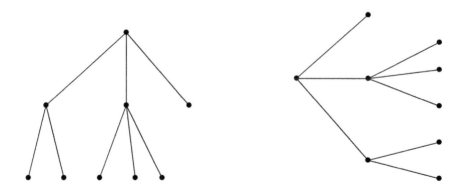

Figure 2.10: Rooted trees may be drawn "down" or "from left to right".

Rooted trees are used to explain and clarify many applications. Integer and other forms of mathematical programming often make use of a branch and bound tree, which is a rooted tree. The tree is used to represent a division of a large problem into two or more parts, and then these parts can be further divided. The start of such a tree is illustrated in Figure 2.11.

Rooted trees are used in subjects other than mathematics to indicate choices or sub-divisions. Figure 2.12 shows a part of the classification of websites used by one organization. The vertices represent a topic, and the directed edges leaving them represent the way that the topic is split into smaller topics. On a simpler level, the index to a book can be considered as a rooted tree, with a vertex for each letter, split into entries in the index, and these may be the head-words for further topics and vertices. The data structure for the files

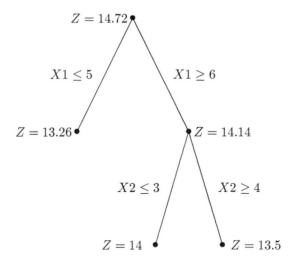

Figure 2.11: Rooted tree used to represent stages in an integer programme: the vertices represent successive relaxed problems, with the objective function value alongside, and the edges show the constraints which are added.

on most computers is a rooted tree, and many programs for accessing files offer the user the chance to see the tree in a diagram. Computer users are familiar with the term **pathname** to locate files. In a rooted tree, it is customary to draw vertices so that they are arranged in levels. The **depth** or **level** of a vertex i is defined as the length of the unique path from the root of the tree to i. The **height** of a tree is the greatest depth of any vertex in the tree.

Because there is a similarity between family trees (showing the ancestors and/or descendants of a person) and rooted trees, some of the names are related. Vertices which are on the same level are called **siblings**. If the edge (i, j) is in the tree, then i is the **parent** of the **child** or **offspring** vertex j. For any vertex v, the path through the parent of v to the root of the tree passes through the many **ancestors** of v.

When there is a limit m on the number of children that a particular vertex may possess, then the tree is called an m-**ary tree**. Binary ($m = 2$) and ternary ($m = 3$) trees are the most commonly encountered. When a binary tree such as Figure 2.11 is drawn, the children may be identified as being left-child and right-child, although which is which is often arbitrary.

Binary trees have a great many applications in computing. Further details can be found in appropriate text-books.

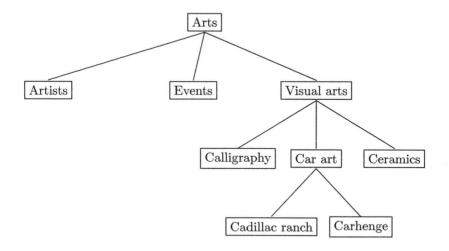

Figure 2.12: A rooted tree used to classify websites

2.4   EXERCISES

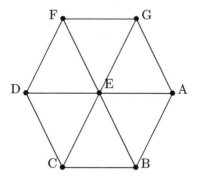

Figure 2.13: First graph for exercises.

1. Find the minimum spanning tree given the matrix of edge weights in
   Table 2.2. Use two different methods and compare the amount of work
   needed.

	1	2	3	4	5	6	7
1	0	517	235	390	547	356	237
2	517	0	282	185	167	172	300
3	235	282	0	184	359	121	59
4	390	185	184	0	177	85	157
5	547	167	359	177	0	249	315
6	356	172	121	85	249	0	147
7	237	300	59	157	315	147	0

Table 2.2: First table of weights for exercises.

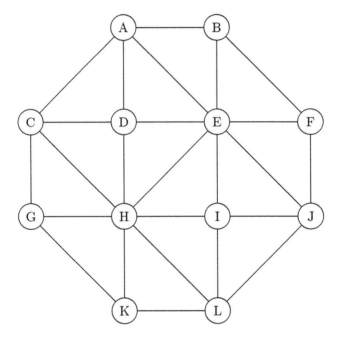

Figure 2.14: Second graph for exercises.

2. Find the minimum spanning tree given the matrix of edge weights in Table 2.3.

3. For many problems with graphs and networks, one can draw one's own diagram, choose a set of numbers and solve the resulting problem. Figure 2.13 is a suitable diagram for this. The data in Table 2.2 can be used with the diagram, as follows. Randomly choose a way of matching the letters {A, B, C, D, E, F, G} to the numbers 1 to 7, use the diagram to

	1	2	3	4	5	6	7	8	9	10	11	12
1	0	52	23	39	55	36	24	51	9	37	57	24
2	52	0	28	18	17	17	30	7	43	18	8	39
3	24	28	0	18	36	12	6	27	15	16	34	11
4	39	18	18	0	18	9	16	13	31	5	22	29
5	55	17	36	18	0	25	32	11	51	22	11	47
6	36	17	12	9	25	0	15	16	27	4	23	23
7	24	30	6	16	32	15	0	26	15	14	33	16
8	51	7	27	13	11	16	26	0	42	14	7	38
9	9	43	15	31	51	27	15	42	0	29	49	15
10	37	18	16	5	22	4	14	14	29	0	21	27
11	57	8	34	21	11	23	33	7	49	21	0	45
12	24	39	11	29	47	23	16	38	15	27	45	0

Table 2.3: Second table of weights for exercises.

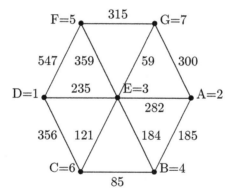

Figure 2.15: One of the hundreds of possible graphs generated by the rule in exercise 3.

determine which edges exist, and the table to obtain the edge weights. Find the minimal spanning tree. (Using a deck of cards, pick out the ace to seven of spades, shuffle them and then the value of the top card is A in the diagram, the next is B. Figure 2.15 shows the edge weights for A=2,B=4,C=6.D=1,E=3,F=5,G=7.)

4. Repeat the procedure from question 3 using Figure 2.14 and Table 2.3. Compare the amount of work needed with two different algorithms.

5. Generate weights for Figure 2.13 in the same way as in exercise 3, and find the maximal weight spanning tree.

6. Generate weights for Figure 2.14 in the same way as in exercise 4, and find the maximal weight spanning tree.

7. Find weighted graphs $G1, G2$ with n vertices for which the following rules in Kruskal's method will *not* find the minimal spanning tree:

   (a) "List the edges of $G1$ in increasing order of weight, and choose the first $n - 1$";

   (b) "List the edges of $G2$ in increasing order of weight, and choose edges until all vertices are connected".

8. In a tree with n vertices, $v_1, v_2, \ldots, v_n$, there are $n - 1$ edges, and so the sum of the degrees of the vertices will be:

$$\sum_{i=1}^{n} d(v_i) = 2n - 2$$

   Given any set of n positive integers $d(v_i)$ whose sum is $2n - 2$, is it possible to find a tree whose vertices have these as their degree? Explain your answer.

9. Suppose that T_1 and T_2 are two spanning trees on a graph G. (They need not be minimal weight trees.) One edge, e_i is chosen from T_1. Show that there is an edge e_j in T_2 such that $T_1 - e_i + e_j$ is a spanning tree on G.

10. How many ancestors can a vertex in a rooted tree of height h possess?

11. Suppose that T is a binary rooted tree with height h. What is the largest possible number of vertices in T? What is the smallest? What are the corresponding values for an m-ary tree?

12. (Computer work.) Write a computer program or spreadsheet which implements one of the minimal spanning tree methods for graphs with up to 12 vertices and 40 edges.

3

Shortest Paths

When you have read this chapter, you should know:

- The importance of shortest path problems in the study of graphs and networks;

- Examples of the range of practical problems which can be modelled as shortest path problems;

- How to find the shortest path through a directed graph.

3.1 INTRODUCTION

On the world-wide-web, there are several sites which offer to help to plan a journey. One British example asks for the postal codes of the starting point and destination, and then supplies the user with the shortest route by road, along with a set of travel directions. There are equivalents for many countries and some which can be used for road journeys which cross international boundaries. Another, which I use when planning a journey by rail, asks for the starting point and destination station, requests the time when I wish to arrive at the destination, and offers a choice of train services and connections which will allow me to complete the journey before the time selected—provided that the trains keep to the timetable. Similar sites exist in many countries, and for air journeys as well as road and rail.

How do these sites work? Quite obviously, they do not store routes between every pair of postal codes, or between every pair of stations at every time of day and on every day of the year. Instead, there is a computer program which tries to find the best route for the journey, using a suitable database. This chapter provides an introduction to methods for finding the best route. This is usually the shortest, cheapest, most reliable or quickest. For convenience, the topic will be referred to as the problem of finding the shortest path, rather than constantly referring to these alternatives.

There are other problems which can be formulated as shortest path problems. These include the knapsack problem, production planning, and scheduling staff for 24-hour rotas.

3.2   PATH AND OTHER NETWORK PROBLEMS.

Shortest path problems are important in the study of networks. First, they are very common in practice; one often wants to find the best route for sending something (me, my car, an email or telephone message) from one place to another. Second, they provide the basis for a number of key ideas in the subject material; by looking at path problems, one can establish foundations for a large variety of other, more complex, network models. Third, it is often necessary to solve a path problem before one can start on other, more advanced, algorithms. Fourth, and finally, shortest path problems are easy to solve.

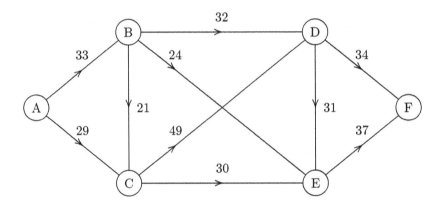

Figure 3.1: A simple shortest path problem. The numbers alongside the edges represent the distances between vertices. What is the shortest path from $s =$A to $t =$F?

3.2.1   Assumptions

Take a weighted graph or digraph, $G = (V, E)$, with weight c_{ij} assigned to each edge (i, j). For any path, the value of the **weight of the path** is the sum of the weights associated with the edges from which it is formed—so that, for example, the measure for the path $i \rightarrow j \rightarrow k$ is $c_{ij} + c_{jk}$. (This assumption doesn't work for fares on public transport.) For the first algorithm, it will be assumed that the edges are undirected and the weight is always a non-negative integer. Figure 3.1 shows a typical small problem. It is customary to identify one vertex as the start of the path, and call it s, and another as the end (or terminus), called t. Given these, the shortest path problem is to find the path of least weight through the graph G from s to t.

3.3   APPLICATIONS

Small problems such as Figure 3.1 can be solved with pencil and paper very quickly. And, generally, finding the shortest distance between two cities is not a problem which recurs—once I know that the shortest distance from my home to that of my brother is 344 kilometres, there is little point in recalculating it unless one of us moves house. The problem needs to be solved once, and probably once only. However, for some of the applications, the weights on the edges of the graph vary, and so it is necessary to find the best path again and again. This section describes how some such problems are modelled using weighted graphs and the shortest path problem.

3.3.1   The knapsack problem

Packing items in a container, such as a suitcase, knapsack or car, frequently involves some choices. Usually, there are more items that could be included than the container will hold. The limit may be because of weight (the airline will not allow more than a certain weight, or a hiker does not want to carry too much) or volume (the suitcase will not stretch beyond its physical limits). How does one choose what to pack, and what to leave out?

This is the problem which is modelled by the **knapsack problem**. The container to be packed will hold, say, L litres. Assume there are N different items which may or may not be included in the packing. The ith of these has a known volume, l_i litres, and a value, v_i euros. Which items should be packed in order that the total volume is less than or equal to L, and the total value is as great as possible? It is assumed that the values are all positive, and that the total volume of all the items is more than L. There are various ways that this problem can be solved in operational research, such as integer programming and dynamic programming. (Trial and error is not recommended. In the worst case, the total volume of each of the 2^N possible sets of items would have to be tested. A brute force approach would have computational complexity $O(2^N)$.) However, as the problem may be modelled as a shortest path problem, that is how it will be described here.

Suppose that $L = 6$ litres, and there are $N = 5$ items. Table 3.1 gives the volumes and values of the items. Suppose also that packing starts with

Item number i	5	4	3	2	1
Volume l_i	4	3	2	2	1
Value v_i	20	17	18	15	7

Table 3.1: Data for a knapsack problem.

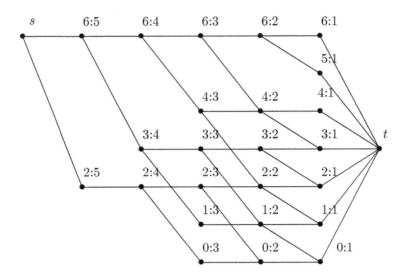

Figure 3.2: The network corresponding to the knapsack problem discussed. Each vertex corresponds to a state of "packing" the container. For clarity, the costs have not been shown; all edges are directed from left to right. The vertices are named $w : i$ where w is the remaining weight in the container, and i is the number of the item which has just been considered. Here s and t are marked.

no items selected, and L litres available. Then, consider the item number 5, decide whether or not to pack it. This will leave some volume available and the next item (number 4) can be considered, and the decision about this will leave some volume available and the next item may be considered, and so on. The numbers are deliberately organized in decreasing order to indicate the way that the decisions follow a "count-down" to zero. The network is modelled with a vertex for each combination of "volume available" and "item number most recently considered" that might occur. Two edges leave most vertices. One corresponds to packing zero items, leaving the same volume available in the knapsack, and this edge has zero weight. The other corresponds to packing one item, which will reduce the volume available, and the item's value will be the edge weight. The problem is to find the most valuable path from the vertex "L available, nothing considered" to a final vertex "all items considered". This is shown in Figure 3.2.

In the figure, the vertices are identified with this two part indication of decisions and state. The edges do not have their value shown (to prevent the diagram becoming cluttered!); the values may be found from the table above. The optimal solution to this knapsack problem is the longest path from vertex s to vertex t. This problem may be converted to a shortest path problem in several ways. The simplest is to note that every path uses 6 edges, and that no edge has a cost greater than 20 euros. So the longest path will have length $P < 120$, and the problem of maximizing P is equivalent to minimizing $120 - P$. If the network is modified by making each cost become $20 - v_i$, all the costs will be non-negative, and the problem will become one of finding the shortest path.

3.3.2 Production planning

In some factories the demand for particular products is known with reasonable accuracy for several time periods (days, weeks or months) in advance. Managers in such factories face the problem of scheduling production, to balance the costs of production against those of storage. To start a production run uses resources of money or time, and then there will be a further cost for each item made. Long production runs may be more efficient, because of the fixed charge, but the result is a surplus of completed goods which need to be stored, and there will be a financial penalty for keeping this surplus in some secure place. It can be proved that, so long as there is no limit to the number of items that can be stored, the optimal production plan only makes items when a time period starts with nothing in storage, and then makes enough for one, two, three or more complete time periods. There are two ways of formulating this problem as a shortest path problem, shown in figures 3.3 and 3.4

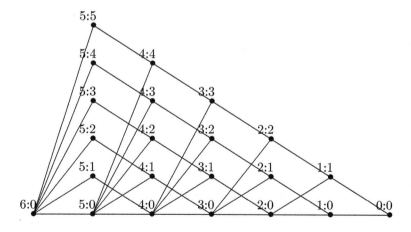

Figure 3.3: The first method of representing a production planning problem. Edges are directed from left to right. The vertices are named $p : i$, where p is the number of time periods remaining, and i is the amount of finished items in storage, measured in units of the number of time periods' supply being held.

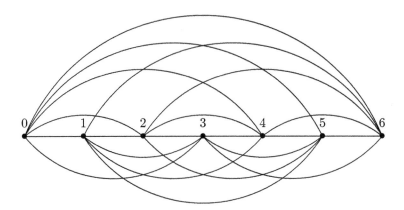

Figure 3.4: An alternative way of representing a production planning problem. Here, each vertex represents a moment in time, the start of one of the periods being considered. An edge connecting two vertices represent the decision to produce sufficient items to supply the demands in the intervening periods; the cost is the cost of setting up a production run, making the items, and storage from one period to the next. A similar network can be used for models of equipment replacement; the edges represent the cost of purchase at one moment in time, maintenance and then sale at the moment in time represented by the final vertex.

3.3.3 Equipment replacement

In the caption to Figure 3.4 there is a note that a similar network can be used for problems of equipment replacement. Businesses which use single items of expensive equipment, or a large number of similar pieces of equipment, find it convenient to consider when to replace them. Once again, it is a matter of balancing two conflicting costs. Keeping a machine for a long period of time reduces the capital cost per year, but the machine will age and require more maintenance. The balance has to be found between capital and running costs, averaged over a long period of time.

When such problems are modelled, one tries to find the expected cost of a decision such as: "purchase at time t_i, sell at time t_j, keep it running all the time in between". This will be the weight (or length) of the edge from vertex t_i to the later vertex t_j. For convenience, one plans for a long time, with the assumption that at the end of the planning horizon, all the equipment can be disposed of. There have been several reports based on this model advising about the optimal policy for replacing family cars. Generally they conclude that it is best to buy a second-hand car which is reliable and which is two to three years old, keep it for a few years and then sell it while it is still reasonably roadworthy. The models assume that there are other people who buy cars which are either new or old and roadworthy, so as to ensure a supply of and a demand for the cars of the optimizers.

3.3.4 Scheduling personnel with changing demand

Organizations which are staffed continually for long periods, such as offices which are open from early morning to late evening, or stores which open at all hours of the day and night face the problem of arranging when people work. This is normally taken to mean deciding on the number of staff who start and finish work at different times. Usually there are suggested minimal numbers of staff needed for different periods of each working day, up to twenty-four hours, and this pattern may be repeated daily, or there may be longer fluctuations. In a supermarket the need for staff to work at check-outs will vary from day to day as well as from hour to hour. Because most employees prefer to have their working hours assigned in spells of reasonable length, scheduling can become complex. Surprisingly, this scheduling problem can be treated as a network, with a shortest path as the objective.

As a simple illustration, consider the problem faced with scheduling eight-hour shifts when the demand for personnel varies according to the data in Table 3.2. Here the staff begin work at midnight, 4am, 8am, 12 noon, 4pm and 8pm. To model this as a network, define $y_i, i = 1, \ldots, 6$ as the number

i	1	2	3
time period	midnight–4am	4am–8am	8am–12 noon
staff needed s_i	9	12	25

i	4	5	6
time period	12 noon–4pm	4pm–8pm	8pm–midnight
staff needed s_i	28	17	13

Table 3.2: Staff needs for each of six four-hour periods during the day. Staff work for two sequential periods, and the manager wishes to find the smallest number of staff needed to satisfy this table of needs.

of staff who start work at each of these six times. Then the numbers working must satisfy the constraints:

$$y_1 + y_2 \geq b_2 \qquad y_2 + y_3 \geq b_3 \qquad y_3 + y_4 \geq b_4$$
$$y_4 + y_5 \geq b_5 \qquad y_5 + y_6 \geq b_6 \qquad y_6 + y_1 \geq b_1$$
$$y_i \geq 0$$

Instead of using the numbers y_i in the model, use the following:

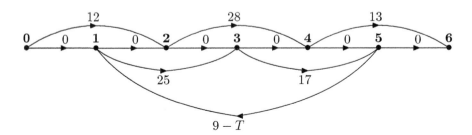

Figure 3.5: The problem of finding the minimum number of people needed for the staffing schedule in Table 3.2 is equivalent to finding the smallest value of T for which there is a longest path, of finite length, from vertex 0 to every other vertex.

$$x_0 = 0, \quad x_1 = y_1, \quad x_2 = y_1 + y_2, \quad x_3 = y_1 + y_2 + y_3$$
$$\ldots \quad x_6 = y_1 + y_2 + y_3 + y_4 + y_5 + y_6 = T$$

with T the number employed. Rewriting the inequalities leads to:

$$x_2 - x_0 \geq b_2 \qquad x_3 - x_1 \geq b_3 \qquad x_4 - x_2 \geq b_4$$
$$x_5 - x_3 \geq b_5 \qquad x_6 - x_4 \geq b_6 \qquad x_1 - x_5 \geq b_1 - T$$
$$x_i - x_{i-1} \geq 0$$

which corresponds to finding the lengths of all the longest paths from node 0 in the network of Figure 3.5. The term T is unknown. For large values of T, there will be a solution, but for small values, the longest paths will repeatedly use the cycle that passes through vertices 1, 2, 3, 4 and 5.

3.4 THE SHORTEST PATH ALGORITHM

This section looks at the most versatile algorithm for finding the shortest path between given vertices s and t.

3.4.1 Informal description

Suppose that the problem has been solved, so that the shortest path from vertex s to vertex t has been found. The shortest path from s to t will use one or more edges and the last one will be incident on t. The vertex which precedes t, say t', in this shortest path, will be closer to s than t, and the length of the path from s to t via t' will be the length of the path from s to t' plus the length of the edge from t' to t. This will be the case, even if $t' = s$ and the shortest path to t is the single edge (s, t). In addition, the part of the path from s to t' must also be the shortest path to t'. Because there is a unique shortest path to any vertex t, the set of edges which are used in all the shortest paths will form a tree.

The algorithm for finding the shortest paths uses these observations. Obviously, the shortest path from s to itself is a trivial one, and has length 0. Then find the vertex that is closest to s, by putting a numerical "label" equal to the length of the edge on all the vertices that are connected to s. Find the smallest of these, and the first interesting (non-trivial) path has been found. Now investigate whether this smallest label could be used as part of a path to other vertices, by seeing if the path that has just been found can be extended. This process is repeated, selecting the path with smallest label that hasn't been used, trying to extend that path, until all the vertices have a path to them.

This algorithm is usually referred to as Dijkstra's ([5]) method. To make it work, there needs to be a simple way of recording which labels definitely represent the lengths of shortest paths, and which might possibly do so. This could be done by recording the most recent path length, but it is more convenient to make the labels have two states, generally called "temporary" and "permanent". These names fit in with another common description of Dijkstra's method as being a "label- setting" algorithm.

3.4.2   Dijkstra's method

The aim is to find the shortest path from vertex s to vertex t with edge lengths given by $d_{ij} \geq 0$ through a network.

At any stage, each vertex, j, has a label, l_j (temporary value), or L_j (permanent value), corresponding to the length of the shortest path that has been found from s to j so far. If this label is permanent, then L_j is the actual length; if not, it may be possible to reduce l_j further in the future. Reduction of a label value happens when there is a path to j which consists of the path to a permanently labelled vertex k followed by the edge (k, j). A label becomes permanent when it is the smallest temporary one left.

The formal statement of Dijkstra's method.

0: Assign a temporary label $l_i = \infty$ to all vertices in the network except vertex s for which $l_s = 0$.

1: (A choice step) Find the vertex k which has a temporary label and for which the label value l_k is least. (Choose arbitrarily in the case of ties; if the tie includes t then this should be chosen.) Make l_k permanent, and call it L_k. If $k = t$, then stop with the length of the shortest path found; the shortest path from s to t is the set of labelled feeder edges, working back from t.

2: (A comparison step) For every vertex j with a temporary label, calculate the smaller of l_j and $L_k + d_{kj}$ and assign this value to the label. If the edge (k, j) is used, label the edge as the feeder for vertex j. If there are temporary labels which are not infinite, go back to step 1, otherwise go to step 3.

3: The vertices with temporary labels all have labels of ∞, so the algorithm must stop; there is no shortest path through the network to t.

3.4.3 Example

Consider the example in Figure 3.1. The tables (3.3 and 3.4) which follow show the progress of the labels, using the algorithm.

Step	labels	k
0	$l_A = 0, l_B = \infty, l_C = \infty,$	
	$l_D = \infty, l_E = \infty, l_F = \infty$	
1	$L_A = 0$	$k = A$
2	$l_B = \min(\infty, 0 + 33), l_C = \min(\infty, 0 + 29),$	
	$l_D = \infty, l_E = \infty, l_F = \infty$	
1	$L_C = 29$	$k = C$
2	$l_B = 33, l_D = \min(\infty, 29 + 49),$	
	$l_E = \min(\infty, 29 + 30), l_F = \infty$	

Table 3.3:

Step	labels	k
1	$L_B = 33$	$k = B$
2	$l_D = \min(78, 33 + 32), l_E = \min(59, 33 + 24),$	
	$l_F = \infty$	
1	$L_E = 57$	$k = E$
2	$l_F = \min(\infty, 57 + 37)$	
1	$L_D = 65$	$k = D$
2	$l_F = \min(94, 65 + 34)$	
1	$L_F = 94$	$k = F$
stop		

Table 3.4:

3.4.4   Trees and Dijkstra's method

There is another way of looking at Dijkstra's method. One thinks about it as a tree-growing algorithm. A **Dijkstra tree** rooted at vertex s is the tree whose edges give the set of shortest paths from s to the other vertices. So the output will be a tree and from it, one can determine the distances. (Earlier, it was noted that this set of paths does form a tree.) As one of these Dijkstra trees is growing, each vertex in the network is either permanently labelled with a distance or temporarily unlabelled. So this is different from the concept above, as there are no distances associated with the unlabelled vertices. Edges which run from labelled vertices to unlabelled ones are known as **frontier edges**. The values of the distance labels correspond to the lengths of the paths from s, as before. The tree starts with vertex s, permanently labelled 0, and then each edge is given a priority equal to the sum of its length and the label at its starting vertex. (So the priority on the edge is the distance to the unlabelled end vertex, using that edge.) The edge with the smallest priority is added to the tree, and the priority is assigned as the label on its terminal vertex.

3.4.5   Example of a Dijkstra tree

With the same example, the diagrams in Figure 3.7 show the growth of the tree of shortest paths for the example, shown in Figure 3.6.

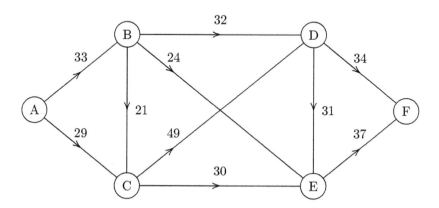

Figure 3.6: Worked example for the shortest path problem. The numbers alongside the edges represent the distances between vertices. What is the shortest path from $s =$ A to $t =$ F?

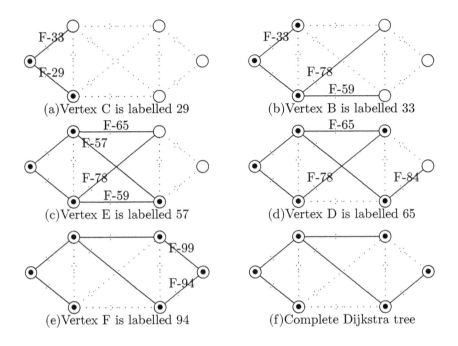

Figure 3.7: Progress growing the Dijkstra tree. (a)—(e) show the frontier edges at each step, and the caption identifies which vertex gains a permanent label. The vertices marked • are those with permanent labels after the step.

3.4.6 Complexity of Dijkstra's method

How much work is needed to find all the shortest paths?

Each pass through the steps of the algorithm makes one more label permanent. So at worst, there will be $|V|$ passes. On the first pass, up to $|V| - 1$ labels have to be examined, on the second it will be $|V| - 2$ and so on, so there will be (in the worst case),

$$(|V| - 1) + (|V| - 2) + \ldots + 2 + 1 = |V|(|V| - 1)/2$$

examinations and calculations, so the complexity is $O(|V|^2)$

3.4.7 Equivalent integer programme for the shortest path problem

Shortest path problems can be formulated as integer programmes. The simplest example is when one wants the shortest path between a pair of vertices, s and t. One way is to choose as the decision variables an indicator to show whether or not an edge is used in a path. (Indicator variables are 0 if something does not happen, 1 if it does.) Then the cost of the path is the sum of the costs of each edge, which will be the sum of the products of the costs on every edge with the corresponding indicators. This will be a linear objective function, suitable for an integer programme. There will be one constraint for each vertex. If the path visits a vertex k, then there must be one edge into k whose indicator is equal to 1, and one edge out of k that is used. So for vertices that are visited, the sum of the indicators for the edges going in must be equal to 1, and so must the sum of the indicators for the edges going out. For vertices that are not visited, the corresponding sums are zero. In each case, the constraint can be written by making the difference of the sums ("in" minus "out") equal to zero. The exception is for vertices s and t. At the start vertex, the difference between the sums will be equal to -1, for vertex t, it will be +1. These constraints are linear equations, so the result will be an integer programme, as below,

$$\min \quad \sum_{ij} d_{ij} x_{ij}$$

$$\sum_{j} x_{sj} = 1 \qquad \sum_{i} x_{it} = 1$$

$$\sum_{i} x_{ik} - \sum_{j} x_{kj} = 0 \quad (k \neq s, t)$$

$$x_{ij} \geq 0 \quad \text{and integer} \quad \forall i, j$$

and the shortest path will have non-zero x-values.

3.4.8   The second shortest path

Finding the shortest path through a network often leads to questions about how good the path is compared with its rivals. One way of answering such questions is to think about the second shortest path, which will differ from the shortest path in at least one edge. The simplest way to find the second shortest path is to:

0: Find the shortest path, which will use a succession of r edges $(s, m_1), (m_1, m_2), \ldots, (m_{(r-1)}, t)$

1: For each of the r edges in turn, fix the edge length to be ∞ and find the shortest path through the modified network. Record the r lengths and pick the smallest one.

This algorithm illustrates the way that many algorithms are building blocks. The shortest path algorithm is used repeatedly as one step here, rather like children assembling several identical components in a construction set.

3.4.9   Equivalent integer programme for second shortest path problem

Assuming that the shortest path has been found, the second shortest path can be found with one integer programme, rather than repeatedly solving path problems in changed networks. The problem below has one extra constraint compared with the shortest path problem; the indicator variables for the shortest path cannot all be 1, so as to ensure that there is at least one edge that is not used in the second shortest path. This constraint can be expressed using the sum of the indicators for the edges on the shortest path, as below,

$$\min \sum_{ij} d_{ij} x_{ij}$$

$$\sum_{j} x_{sj} = 1 \qquad \sum_{i} x_{it} = 1$$

$$\sum_{i} x_{ik} - \sum_{j} x_{kj} = 0 \quad (k \neq s, t)$$

$$x_{sm_1} + x_{m_1 m_2} + \ldots + x_{m_{(r-1)} t} \leq r - 1$$

$$x_{ij} \geq 0 \quad \forall i, j$$

and the second shortest path will again have non-zero x-values.

3.4.10   Sensitivity analysis

What happens if ... the length of an edge in the network is changed? The output for Dijkstra's method is the length of the path and the edges used. So ... the situation is similar to that for the minimal spanning tree.

Assume that the shortest path is a set of edges, P, of length L and edge (i, j) has its length d_{ij} changed; this may affect P; it may affect L. There are four cases considered in Table 3.5. The length of the edge can be increased or decreased, and the edge may or may not be in the path P.

	$(i,j) \in P.$	$(i,j) \notin P.$
Increase d_{ij}.	(1) At first L increases, and P not changed. Then a new path becomes optimal, not using (i,j). Then (2) applies.	(2) Nothing happens.
Decrease d_{ij}.	(3) L decreases, P is not changed.	(4) No change to P, until (perhaps) d_{ij} becomes small enough for (i,j) to be in the shortest path. Then (3).

Table 3.5: Sensitivity analysis for the length of the shortest path, and the edges in it, as the length of edge (i,j) is changed.

3.5 OBVIOUS AND IMPORTANT EXTENSIONS

Dijkstra's method finds the **best** path, so long as one is able to compare two paths and decide which is the better. The rules inside the algorithm need to be changed, and the initial values of the labels will alter, but the essential features remain the same. If one returns to the formal statement, the key features there are two steps.

One is the **choice step** which makes a decision about which label is to make permanent. The other is the **comparison step** which compares a measure of performance associated with two different paths, and selects the one which is preferred. In the case of shortest paths, the choice step is to select the vertex with the smallest label which is still temporary, and the comparison step uses the lengths of the two paths. However ...

... if one was searching for the path with the largest capacity, then the steps, and the labels would be different. The choice step would make the largest temporary label permanent, and the comparison step would choose the path with the larger capacity.

... if one was looking at a bus or train schedule, then the comparison step would be based on the time of arrival, taking into account the time spent waiting for connections.

... and so on.

3.6 EXERCISES

1. Find the shortest path from vertex A to vertex I through the weighted network in Figure 3.8. How would you find the shortest path between these two vertices which visited a specific vertex, such as D?

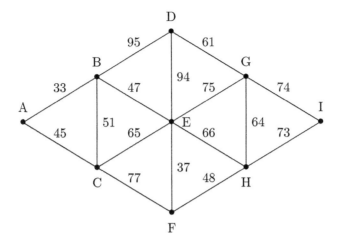

Figure 3.8: The numbers alongside each edge represent the edge length in kilometres.

2. Logician Kurt Gödel revolutionized mathematics with his incompleteness theorem in 1931. This is a puzzle in Douglas R. Hofstadter's Pulitzer prize-winning book "Gödel, Escher, Bach: An Eternal Golden Braid", which Hofstadter uses to introduce the idea of a formal system of logic, which is the basis of Gödel's theorem. (This problem was posed by Scott Kim http://www.scottkim.com/)

   Start with the string of letters MI. Strings can be changed in four ways:

   (a) If a string ends with I, you can add a U at the end. MI can become MIU.

   (b) Any string Mx, where x stands for any string of letters, can be turned into Mxx. For instance, MIU can become MIUIU, and MUM can become MUMUM.

   (c) If there are three consecutive I's in a string, they can be replaced with one U. MIII can become MU. MUIIII can become MUUI or MUIU.

   (d) If there are two consecutive U's in a string, you can drop them. For instance, MUUUIII can turn into MUIII. MUUIUU can become MI.

   The challenge is to make each of the following strings, starting with MI, in the fewest possible steps. None of the possible strings takes more than 10 steps but two of them are impossible to make.

   (a) MUI
   (b) MUIMUI

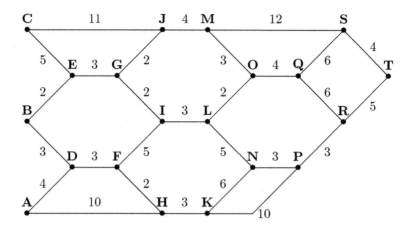

Figure 3.9: On the road; journey times in hours

(c) MIUU

(d) MUIUIU

(e) MIUUIIUUI

(f) MIIIIIIII

(g) MIIIIIII

(h) MIIIIII

The network approach is: Create a network where each vertex represents a possible string; the directed edges of length 1 represent a way of transforming the string. Now find the shortest path through the network from the start vertex labelled MI.

3. I was quietly singing "I am only twenty-four hours from Tulsa" and my road atlas of the USA gave me the data shown in Figure 3.9. Alongside each road section is the time of travel in hours obeying speed limits on the roads. Tulsa is marked as "T" in the figure. Where was I, assuming that I would travel to Tulsa by the quickest route?

4. As in the chapter on spanning trees, it is relatively easy to draw a graph, pick some weights and find the shortest path as a way of gaining practice with the algorithms. Figure 3.10 gives a digraph, and Table 3.6 give the lengths of the edges, once a way of randomly assigning the letters to the numbers has been chosen. Make such a random choice, and find the shortest path from C to J.

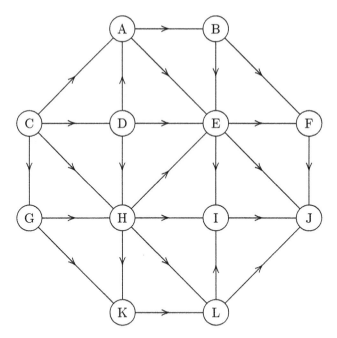

Figure 3.10: Directed network for exercises

5. $s \neq i$ is a vertex in an undirected graph G with positive weights on the edges, and the edge (i, j) is the shortest edge in the graph. Prove, or find a counter-example to the claim: "There must be at least one shortest path from s which includes edge (i, j)". What happens when $s = i$?

6. Give an example of a graph G where the longest edge (k, l) is in at least one shortest path from vertex s.

7. For a given destination vertex t in a network with edges E, the "most vital edge" is defined as that whose removal from E will increase the optimal path length from s to t by the greatest amount. This path consists of the set of edges P. Which of the following statements are true, and which are false? For a statement which is true, justify your answer; for one which is false, give a simple counter-example.

   (a) The most vital edge for a given vertex t is unique.
   (b) The most vital edge is the edge in E whose length is greatest.
   (c) The most vital edge is the edge in P whose length is greatest.
   (d) The most vital edge could be the edge in P whose length is least.

8. (Computer work.) Write a computer program or spreadsheet which implements Dijkstra's method for graphs with up to 12 vertices and 40 edges.

	1	2	3	4	5	6	7	8	9	10	11	12
1	0	52	23	39	55	36	24	51	9	37	57	24
2	52	0	28	18	17	17	30	7	43	18	8	39
3	24	28	0	18	36	12	6	27	15	16	34	11
4	39	18	18	0	18	9	16	13	31	5	22	29
5	55	17	36	18	0	25	32	11	51	22	11	47
6	36	17	12	9	25	0	15	16	27	4	23	23
7	24	30	6	16	32	15	0	26	15	14	33	16
8	51	7	27	13	11	16	26	0	42	14	7	38
9	9	43	15	31	51	27	15	42	0	29	49	15
10	37	18	16	5	22	4	14	14	29	0	21	27
11	57	8	34	21	11	23	33	7	49	21	0	45
12	24	39	11	29	47	23	16	38	15	27	45	0

Table 3.6: Second table of weights for exercises.

9. Lewis Carroll, the famous Victorian mathematician and pioneer photog-
rapher, is considered to be the originator of the puzzles known as "word
ladders". These comprise a starting word of n letters and a finishing
word of the same length; both must be recognized words of some kind
(e.g from a dictionary or place names). The aim of the puzzle is to
change the starting word into the finishing word, by altering one letter
at a time, with every set of n letters forming a word. Having done that,
one tries to do it as quickly as possible. Thus, to change "**BOY**" into
"**MAN**", one could create a ladder:

B	O	Y
B	A	Y
M	A	Y
M	A	N

(a) Why is this exercise a problem in shortest paths?

(b) How do you change "MAN" into "BOY" without using the same
intermediate words?

(c) And "CAT" into "DOG" and back again, using all different words?

(d) Finally, "WALK" into "PATH"?

4

Maximum Flows

When you have read this chapter, you should know:

- The place of the maximum flow problem in the study of networks;

- Examples of practical problems which can be modelled as maximum flow problems;

- How to find the largest flow through a capacitated graph from one or more sources to one or more destinations;

- The connection between flows and cuts.

4.1 INTRODUCTION

Each edge (i, j) in a directed network has a maximal capacity u_{ij} for flow of some item or material, measured in convenient rates per unit time. So it may be per hour, per day, or even per year. At the vertices, the flows in edges can be split into parts, or combined, so long as the total flow is conserved at each vertex. All the flow into a vertex must leave it, except at the start and finish (destination). Therefore there are constraints on the edges and at the vertices. Given a start vertex s and a destination t, the problem is to find the maximum possible flow from s to t which conserves flow and does not violate any of the capacity constraints.

4.1.1 Why is it important?

The obvious application of this method is to finding the largest flow between a source of goods (or several sources, which can be combined into a super-source) and a destination for them. So it applies to problems of supplying gas, water, electric power and communication transmissions, and with simple variations, to road, rail and maritime transport. There have been examples in the operational research literature of the problems of moving people to and

site	cost of site	Cost of link to:			
		A	B	C	D
A	40	×	12	42	39
B	40		×	12	15
C	40			×	45
D	40				×

Table 4.1: The costs of the sites and the revenues for each link for the example of Balinski's site location problem.

from sporting events, and in planning the evacuation routes around nuclear power plants. In the same way as the shortest path problem is a part of larger problems of networks, so the maximum flow problem is often part of many larger problems, such as arranging road transport schedules over several days or weeks.

4.1.2 Balinski's site selection.

In 1970, Balinski[1] showed that the following problem can be written as a maximum flow problem. There are n possible sites for a business to use. If site i is used, then there will be a cost c_i per year for running it. However, if sites i and j are both developed, then there will be an annual income of k_{ij} from the connection between them. This might be the case, for instance, if the two sites communicate with each other in some way; this is a simplified model of electronic communications links. Which sites should be operated in order that the profit is made as large as possible?

This is an example where the problem has a network structure which must be modified in order that the problem can be solved. Suppose that there are only four sites, with costs and income as shown in Table 4.1.2. The problem can be pictured by the network in Figure 4.1. The optimal configuration will use no sites, two sites, three sites or all four sites. For this small example the solution could be found by finding the total cost of every possible configuration (there are 12 for this example), but realistic sized problems will have too many configurations for this approach to work successfully. The paper by Balinski shows that the problem of deciding which sites to use can be rewritten so that every site and every link is represented by a vertex in a "logical network". The vertices representing sites are connected to the vertices for the corresponding links by edges whose capacity is infinite. There is a source for a logical flow of material connected to each the vertices representing the links, with an edge of capacity equal to the revenue for the link, and a similar edge between the vertices representing the sites to a destination for the flow. The solution to the site selection problem is a way of cutting the network between the source and destination for the logical flow. The vertices on the same side of the cut as the

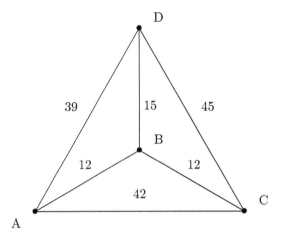

Figure 4.1: The four sites each cost 40 units per year. The revenue from the links between sites varies, as shown alongside the edges.

source will be the ones selected, and this set of vertices will include both the sites and their links. It turns out that the solution is the same cut as is found by the maximum flow algorithm. Figure 4.1.2 shows the logical network, with the source s and destination t marked.

4.1.3 Planning open-pit mining

A further application of the maximum flow algorithm is planning where an open-pit (or open-cast) mine or quarry should expand. Blocks of stone and earth are removed and processed to obtain the mineral ores that they contain. One cannot remove a block until there is clear access to it, meaning that the block(s) directly above it have been removed. Additionally, safety constraints mean that the slope of the mine must be gentle, so access to a block that is deep below the surface means that several blocks not directly above it must have been removed. A detailed description of the formulation as a maximum flow problem is given by Hochbaum and Chen ([12]).

4.1.4 A typical small problem.

The example in Figure 4.1.4 shows a typical small problem. How much material can be sent from s to t without violating the constraints? If one examines vertices 1 and 2. some of the complexities of the problem start to become apparent. If edge $(s, 1)$ is filled, then the edges out of vertex 1 will be overfull; so edge $(s, 1)$ cannot be filled. Filling edge $(1, 2)$ will mean that edge $(s, 2)$ cannot be filled. And so on; there is considerable interaction between the flows in edges.

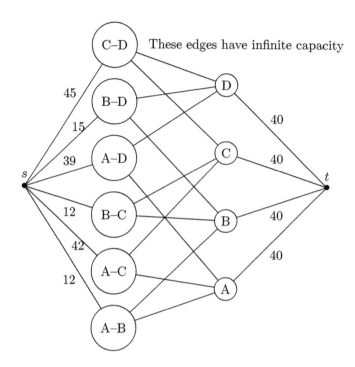

Figure 4.2: The logical network corresponding to the site selection problem of Table 4.1.2 and Figure 4.1. The optimal solution uses three sites, A, C and D.

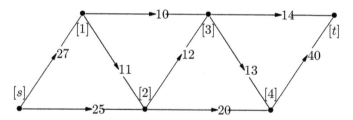

Figure 4.3: A network with an upper bound on the flows. The *Maximum flow problem* is to find the largest flow possible from s to t, and the flows in each edge which gives this.

4.2   FORD-FULKERSON METHOD

In this chapter, the presentation is concerned with a method which is intuitively simple, while guaranteeing to find the optimal solution. It is due to Ford and Fulkerson [10]

Suppose that a set of flows has been found that satisfy the constraints at each vertex, and in each edge. The total flow can then be measured by summing the flows out of the start vertex (s) and subtracting any flows which go into it; equivalently, one can sum the flows into vertex t and subtract any flow which goes out. But is this total flow as large as possible? If it isn't, then there must be some way of sending extra flow from s to t, without breaking any of the capacity constraints.

The algorithm works by finding a succession of flow-augmenting chains (FACs), which are chains from the source to the destination along which some extra flow can be sent. So, given any set of feasible flows, the algorithm has the outline below.

Find a *chain* from s to t with spare capacity.

Send as much material as possible along this chain.

Note the flow that is used in each edge.

Start again.

In order for this to work, it needs two further elements. First, there should be a rule for finding a suitable chain. Second, there will nave to be a stopping rule.

These two elements will be combined; Ford and Fulkerson's method is one which systematically tries to find chains for sending extra flow, and if this is not successful, then it stops, and there is a guarantee that the largest flow has been found.

4.2.1   Formulation of the maximum flow problem.

The maximum flow problem is another linear programming problem because it can be formulated with variables x_{ij} representing the flow in each edge (i, j). There are upper bounds on the flow in each edge, in the form: $0 \le x_{ij} \le u_{ij}$. There are flow conservation equations for all vertices j except s and t, so that: $\sum_i x_{ij} - \sum_k x_{jk} = 0$. The flow out of s or into t gives the flow "through" the network, so that the objective is to maximise $\sum_k x_{sk} - \sum_i x_{is}$. Hence the objective is linear, and all the constraints are linear, and the variables are non-negative. But a solution using the network algorithm is much more convenient and efficient than using the simplex method.

It must start with a feasible flow (which means that the flow is conserved at every vertex, except the source and destination) and look for chains which have some extra capacity.

Usually, this means that the algorithm will try and fill edges up, but sometimes there has been a "mistake" which has to be put right by reducing the flow and redirecting it. So the chains usually have edges which are forward, in the sense of going from source to the destination, but reverse edges can also be useful. Figure 4.2.1 illustrates the problem where flow must be redirected. Clearly, two units of flow can be sent through the network from s to

t. However, no extra flow can be sent from s to vertex 1, as this edge is full. If a unit of flow is sent from s to vertex 2, then the edge $(2, t)$ is full. The problem occurs because edge $(1, 2)$ has been used, and the flow is blocking further progress. The algorithm of Ford and Fulkerson provides a simple way

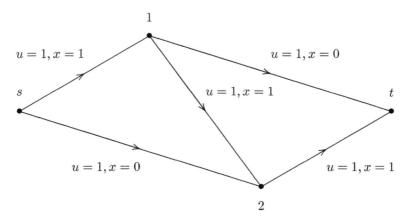

Figure 4.4: In this network, the largest possible flow will be 2 units from s to t. However, the flow of one unit in the edge $(1, 2)$ blocks an increase of flow, illustrating the need for the concept of reverse edges in flow-augmenting chains.

of handling such situations. Flow-augmenting-chains are made up of a mix of "forward" edges, which have spare capacity, and which point in the direction "from s to t" and "reverse" edges, which point in the other direction and which have some flow which can be reduced. In the example of Figure 4.2.1, the flow in edge $(1, 2)$ can be reduced by one unit, and a flow augmenting chain can be defined as edge $(s, 2)$ (forward), edge $(1, 2)$ (reverse) and edge $(1, t)$ (forward). The capacity of this chain for extra flow is one unit, which is the largest change possible in any of the edges without violating a constraint. The constraints are the unused capacity for forward edges, and the existing flow for the reverse edges.

With this new flow-augmenting chain, the pattern of flows in the network will become that seen in Figure 4.2.1.

4.2.2 A formal statement of the algorithm for the maximum flow through a network

Having discussed the algorithm informally, what follows is a formal statement of the steps of the method, including the way that the vertices are labelled to make it easier to find the capacity of a flow-augmenting-chain. The aim is to find the largest flow from vertex s to vertex t with capacities u_{ij} on edges

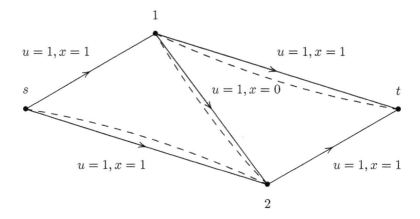

Figure 4.5: The flow in the network of Figure 4.2.1 after one unit of flow has been sent along the flow-augmenting chain (shown as dashes) $(s, 2)$ (forward), edge $(1, 2)$ (reverse), edge $(1, t)$ (forward). Flow is increased in forward edges, reduced in reverse edges.

$(i, j))$

  0: [Initialization] Give each edge (i, j) a flow x_{ij} which is feasible, ensuring that flow is conserved at all vertices (with the possible exception of vertices s and t) This *may* be done by making all flows zero.

  1: Label vertex s with the label $(-, \infty)$ and ensure that no other vertex is labelled. The first part of the label is the preceding vertex in the current *flow-augmenting chain* and the second is the amount of material which can be sent from s to the selected vertex.

  2: [try to extend any part-formed FACs.] Scan through the edges (the exact method is not important) until one (i, j) is found for which either:

    2a: vertex i is labelled and vertex j is not and

$$x_{ij} < u_{ij} \qquad \text{(forward edge)}$$

    2b: vertex j is labelled and vertex i is not and

$$x_{ij} > 0 \qquad \text{(reverse edge)}$$

  If there is no such edge, then go to step 5.

  3: [extend one part-formed FAC, by adding another edge to it.] If the edge found is a forward one, then label vertex j with the two-part label

$(a_j, b_j) = (i, \min(b_i, u_{ij} - x_{ij}))$; if it is a reverse edge, then label vertex i with the two-part label $(a_i, b_i) = ((j, \min(b_j, x_{ij}))$; if vertex t is labelled, then do step 4 otherwise return to step 2.

4: [a flow augmenting chain (FAC) has been found] Increase the flow in the edges which form the chain by the amount b_t. The chain can be found by searching backwards from vertex t using the first parts of the labels to determine the preceding vertex. (in practical implementations, it is useful to be able to determine whether an edge has been used as a forward edge or reverse edge in the chain; this can be done by setting the first part of the label as being positive or negative.) Go to step 1.

5: The optimal flow has been found, so stop.

Step 2 is the crucial step for finding the FAC. In a way which has similarities to tree-growing algorithms, it works by extending chains from labelled vertices to unlabelled ones. However, the labels here are in two parts, identifying the previous vertex and the capacity of the flow-augmenting-chain to the current vertex.

Each time that flow is increased along a FAC, one edge acts as the bottleneck; either one (at least) of the forward edges is filled or one (at least) of the reverse edges is given zero flow. The capacity of the flow-augmenting-chain is fixed by the amount that may be sent through the bottleneck. Ideally, one would like the capacities of bottlenecks to be as large as possible, but there is no easy way to find them in advance.

When the algorithm stops, it is because no edges can be found in step 2. Vertex s will have a label, and so may other vertices. Vertex t will not be labelled, and there may be other unlabelled vertices. The edges which link between the labelled and unlabelled vertices will be those which could not be used to extend any of the part-formed FACs. So the forward edges are all full, and the reverse edges are empty.

4.2.3 The complexity of this algorithm.

Usually, this algorithm is very quick and efficient; however, in theory it might be very slow, because the worst case complexity with all the edge capacities integers is $O(|N||A|U)$ where U is the ratio of the capacity of the largest edge in the network to the smallest. The problem is caused (in part) by the arbitrary nature of the choice of edges and vertices at several stages of the method.

One can deduce the complexity by looking at an absurd example, shown in Figure 4.6. In this network, it is obvious that the maximum flow is U or 198 units, whichever is the smaller, but ... the algorithm may be slow to find this. Suppose that the first flow-augmenting chain found is the one which goes $s \rightarrow 1 \rightarrow 2 \rightarrow 3 \rightarrow t$ with capacity 1 unit, and then the next is the chain $s \rightarrow 2 \rightarrow 1 \rightarrow 3 \rightarrow t$ with capacity 1 unit. Then there will be U or 198 successive flow-augmenting-chains found, and the number of increments is

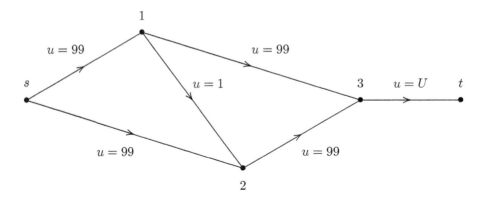

Figure 4.6: The worst-case complexity of the max-flow algorithm. It is possible to find FACs as follows: $s \rightarrow 1 \rightarrow 2 \rightarrow 3 \rightarrow t$ (1 unit capacity); $s \rightarrow 2 \rightarrow 1 \rightarrow 3 \rightarrow t$ (1 unit capacity); . . . and so on . . . .

proportional to the largest capacity in the network. To find each chain, one could need to examine every vertex, and every edge in the network for every vertex, giving the number above. Admittedly, this is an absurd example, but the figure for the complexity has to reflect this worst case situation.

4.2.4   The maximum-flow, minimum-cut, theorem

The algorithm will come to a halt when it is no longer possible to label any vertices, and the destination is left unlabelled. It is intuitively obvious that the flow in the network is now maximum. This is demonstrated by the **Maximum-flow-minimum-cut theorem**.

Theorem 3 Maximum-flow-minimum-cut theorem. *The largest possible flow between vertices s and t in a network is equal to the smallest capacity of any of the cuts that separate s and t.*

Proof: The capacity of any cut $\{X, \overline{(X)}\}$ is equal to $C(X) = \sum_{i \in X, j \in \bar{X}} u(i,j)$. The flow from s to t will be less than or equal to the flow from X to $\overline{(X)}$, as it is possible that there could be flow from $\overline{(X)}$ to X. In turn the flow from X to $\overline{(X)}$ is less than or equal to $C(X)$. So for any cut X, the maximum flow from s to t is less than or equal to $C(X)$.

Ford and Fulkerson's method has stopped with a cut $\{L, \overline{(L)}\}$ defined by the labelled vertices L. For all the edges from L to $\overline{(L)}$, the flow equals the capacity, so the total flow is equal to $C(L)$. Further, there is zero flow from $\overline{(L)}$ to L. Hence there is a cut whose capacity equals the flow when the algorithm stops. This capacity must be the smallest of all possible cut capacities, and the theorem is proved.

	Vertex				
Iteration	s	1	2	3	t
1	$(-,\infty)$				
1	$(-,\infty)$	$(s,30)$			
1	$(-,\infty)$	$(s,30)$	$(1,19)$		
1	$(-,\infty)$	$(s,30)$	$(1,19)$	$(2,12)$	
1	$(-,\infty)$	$(s,30)$	$(1,19)$	$(2,12)$	$(3,12)$

Table 4.2: The labels on vertices for the first FAC in the network.

4.2.5 Worked example

As an example, consider the network shown in Figure 4.7 and the problem of finding the largest flow from the source to the destination. In order to

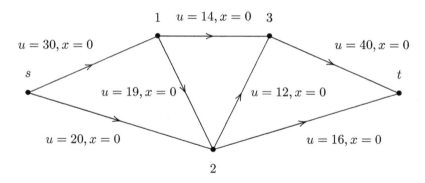

Figure 4.7: What is the largest flow from s to t in this network?

demonstrate the use of reverse edges, the FACs are not found in what might be considered the obvious order. However, as has been observed earlier, the sequence in which edges and vertices are examined is arbitrary; if one did not have a diagram of the network, it would be harder to decide which is an ideal sequence for examining the constituent parts of the network when trying to find the next vertex to label.

Table 4.2 shows the labels which correspond to the first FAC, which has capacity 12 units, and is shown by the dashed lines in Figure 4.2.5.

Successive tables and figures show how the algorithm labels the vertices and changes the flows. In Table 4.3, the edge $(1,2)$ is used as a reverse edge. At this iteration, the reverse edge sets the limit for the capacity of the FAC, since the flow $x(1,2)$ cannot become negative.

| Iteration | s | Vertex | | | |
		1	2	3	t
2	$(-,\infty)$				
2	$(-,\infty)$		$(s,20)$		
2	$(-,\infty)$	$(1,12)$	$(s,20)$		
2	$(-,\infty)$	$(1,12)$	$(s,20)$	$(1,12)$	
2	$(-,\infty)$	$(1,12)$	$(s,20)$	$(1,12)$	$(3,12)$

Table 4.3: The labels on vertices for the second FAC in the network. Note the reverse edge.

| Iteration | s | Vertex | | | |
		1	2	3	t
3	$(-,\infty)$				
3	$(-,\infty)$	$(s,18)$			
3	$(-,\infty)$	$(s,18)$		$(1,3)$	
3	$(-,\infty)$	$(s,18)$	$(1,18)$	$(1,3)$	
3	$(-,\infty)$	$(s,18)$	$(1,18)$	$(1,3)$	$(3,2)$

Table 4.4: The labels on vertices for the third FAC in the network.

| Iteration | s | Vertex | | | |
		1	2	3	t
4	$(-,\infty)$				
4	$(-,\infty)$	$(s,8)$			
4	$(-,\infty)$	$(s,18)$			$(2,8)$

Table 4.5: The labels on vertices for the fourth FAC in the network.

| Iteration | s | Vertex | | | |
		1	2	3	t
5	$(-,\infty)$				
5	$(-,\infty)$	$(s,16)$			
5	$(-,\infty)$	$(s,16)$	$(1,16)$		
5	$(-,\infty)$	$(s,16)$	$(1,16)$		$(2,8)$

Table 4.6: The labels on vertices for the fifth FAC in the network.

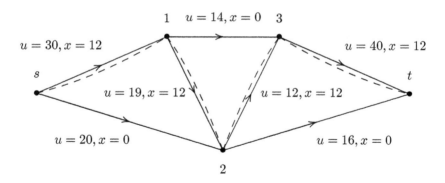

Figure 4.8: The flow in the network after the first FAC has been found; the dashed lines show the FAC.

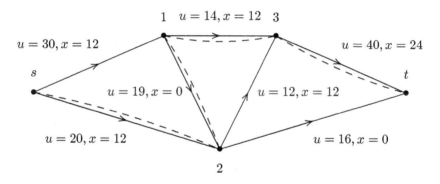

Figure 4.9: The flow in the network after the second FAC has been found; the dashed lines show the FAC.

In the final iteration, the labels are created with values shown in Figure 4.7. It is not possible to label vertex t. The set of labelled vertices, $L = \{s, 1, 2\}$ defines the cut of minimal capacity, the edges $\{(1, 3), (2, 3), (2, t)\}$, for which the total is 42 units. Examination of the network shows that every other cut has capacity at least 42. (Table 4.2.5)

4.2.6   Sensitivity analysis

In any edges that are in minimal-cuts of the network, the optimal flow must be equal to the capacity. In other edges, the flow that is found by the Ford and Fulkerson algorithm need not be the only one that is possible. Flows can be rearranged in edges which are not part of the bottleneck between s

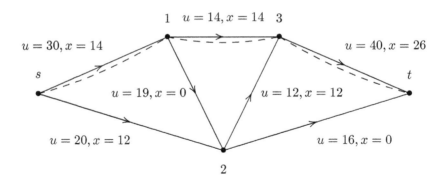

Figure 4.10: The third FAC only increases the flow by 2 units.

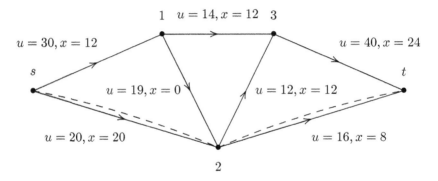

Figure 4.11: The flow in the network after the fourth FAC has been found; the dashed lines show the FAC.

and t. In the example that has just been examined, the flow in edge $(3, t)$ is fixed, even though this is not part of a minimal-cut. But the flows in the edges $(s, 1), (s, 2), (1, 2)$ can be changed, so long as exactly 42 units flows out of vertex s, at least 12 units flows into vertex 1 and exactly 28 units flows into vertex 2. This might be useful knowledge if one wished to rearrange the flows for any reason.

In general, the maximum flow is most sensitive to the parameters of the minimal cut-sets. It is possible to summarize the sensitivity of the value of the maximum flow in a network by means of a table (Table 4.9). Here the parameter that is being changed is the capacity of an edge, and the table shows the effects depending on whether the edge is or is not in a minimal cut.

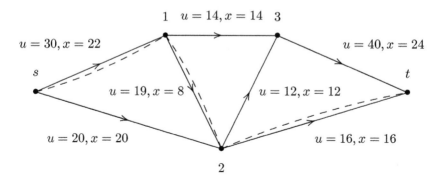

Figure 4.12: The flow in the network after the fifth FAC has been found; the dashed lines show the FAC.

		Vertex			
Iteration	s	1	2	3	t
6	$(-,\infty)$				
6	$(-,\infty)$	$(s,8)$			
6	$(-,\infty)$	$(s,8)$	$(1,8)$		

Table 4.7: The algorithm stops on the sixth iteration. No other vertices can be labelled.

Set containing s	Total capacity
$\{s\}$	50
$\{s,1\}$	53
$\{s,2\}$	58
$\{s,3\}$	90
$\{s,1,2\}$	42
$\{s,1,3\}$	79
$\{s,2,3\}$	86
$\{s,1,2,3\}$	56

Table 4.8: The capacity of all the cuts that separate vertex s from vertex t.

	edge in a minimal cut	edge not in a minimal cut
Increase capacity.	At first the flow increases. Then a new cut becomes minimal and there is no change in the flow.	Nothing happens.
Decrease capacity.	The flow decreases.	No change to total flow (but it may be redirected) until the capacity is small enough for the edge to be in a minimal cut, and then one is back to the minimal cut case.

Table 4.9: Sensitivity for a maximum flow problem.

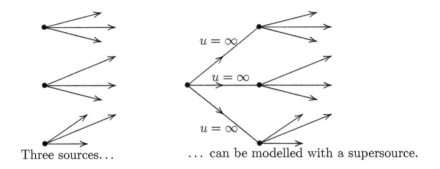

Three sources...   ... can be modelled with a supersource.

Figure 4.13: A supersource

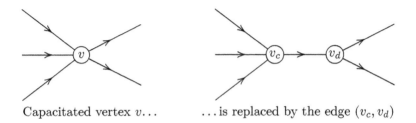

Capacitated vertex v...   ...is replaced by the edge (v_c, v_d)

Figure 4.14: A vertex with constrained flow.

4.3 MULTIPLE SOURCES AND DESTINATIONS

There are often problems where one wants the largest flow from several sources and/or the flow to several destinations. The sources might be several drilling rigs supplying crude oil through a network to a central terminal; the destinations might be several consumers of water, supplied from one reservoir. Such problems can be modelled using a super-vertex (a supersource or supersink) to draw together the separate sources and/or destinations. Figure 4.13 shows the transformation of three sources to a problem with a single supersource, linked to the three by edges whose capacity is infinite.

4.4 CONSTRAINED FLOW THROUGH A VERTEX

In some models, there are limits on the flow through one or more vertices. Such problems are dealt with by treating each such **capacitated vertex** as two vertices, joined by an edge whose capacity is the flow limit. The first vertex acts as the collector of flow and the second as a distributor. All the edges directed towards the original vertex are directed towards the collector; all the edges which were directed away are directed away from the distributor, so that all flow through the original vertex must flow through the edge between collector and distributor. This is illustrated in Figure 4.14.

4.5 EXERCISES

1. Three couples (husband and wife) are going for a meal together. The table has three seats on each side. The couples agree that no husband and wife will sit opposite each other. Create a network with vertices for each pair of seats and each couple, with edges between them to indicate that a member of the couple is or is not in a seat. Show that an allocation of the seats for the meal can be found by solving a maximum flow problem on this network, with the addition of a "super-source" and a "super-sink".

2. Generalize the situation described in the previous question for F families, with $m(i)$ members in the ith, going for a meal at a restaurant with T tables seating 10 people, so that there are no more than two members of each family at any table.

3. In Figure 4.15, how does the maximum flow from s to t vary as X and Y change in the range (0,35)?

4. Find the maximum possible flow from Metz to Paris given the data in Table 4.10.

5. A capacitated network has directed edges, each of which has an associated positive capacity, u_{ij} units of flow per unit time. The Ford-Fulkerson algorithm is used to find the maximum flow through this net-

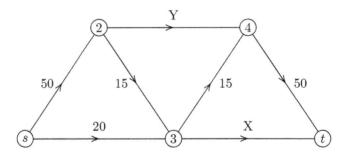

Figure 4.15:

	Met	Mon	Mul	Nan	Nan	Nic	Nîm	Orl	Par
Metz		80	45	50					
Mon'				70	75				
Mul'				30	85				
Nancy					67	87	65		
Nantes						100	85		
Nice							40	95	90
Nîmes								65	80
Orléans									120

Table 4.10: Capacities of routes in France. Mon' is Montpellier, Mul' is Mulhouse.

work, from a source s to a destination t. This is F_{max} with corresponding flows in edges x_{ij}, One edge, (k, l) is changed to become undirected. How will this change affect F_{max} if:

(a) x_{kl} was zero in the optimal solution?

(b) x_{kl} was nonzero in the optimal solution?

6. How can one find the maximum flow through an *undirected* network, where each edge has an upper bound on flow?

5

How to Store a Network

When you have finished this chapter, then:

- You will recognize the most commonly used forms of data storage for networks and graphs, when diagrams are not appropriate;

- You will know that algorithms are helped by the form of the data storage.

5.1 INTRODUCTION

Since many of the algorithms being discussed will be implemented on a computer, it is sensible to think about the way that a network can be stored in a file or as a numerical record. Two uses are important; the user—or the algorithm—will want to access parameters the data and (possibly) to change the values of the variables. Good methods of storage can accelerate the algorithms, bad ones can slow them down.

Two kinds of information must be recorded for most of the algorithms:

1. The topology of the network, that is the underlying graph structure, of what vertices there are, and what edges exist;

2. The parameters of the edges, and any variables which are needed for a given problem.

In Figure 5.1 there is a small network with two parameters on each edge. How can one store this network in a computer for ease of use by the sort of algorithms that are being considered? In standard notation, the graph is:

$$G = (V, E)$$
$$N = \{1, 2, 3, 4, 5\}$$
$$A = \{(1, 2), (1, 3), (2, 3), (2, 4), (3, 4), (3, 5), (4, 5), (5, 4)\}$$

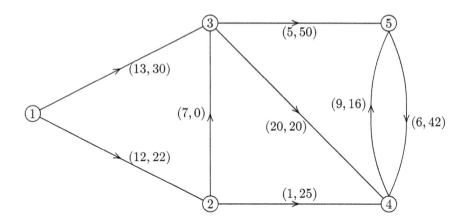

Figure 5.1: A simple network (example 1) with two parameters on each edge.

5.2 VERTEX-EDGE INCIDENCE MATRIX

The first two methods of storage concentrate on the underlying structure of the graph. The vertex-edge incidence matrix (or simply the incidence matrix) for a network has one row for each vertex and one column for each edge of the underlying graph. The column which corresponds to a given edge (i, j) has two non-zero entries, +1 in row i, -1 in row j. Thus for the example, the incidence matrix is in Figure 5.2. The result is a very sparse matrix with most

	$(1,2)$	$(1,3)$	$(2,3)$	$(2,4)$	$(3,4)$	$(3,5)$	$(4,5)$	$(5,4)$
1	1	1	0	0	0	0	0	0
2	-1	0	1	1	0	0	0	0
3	0	-1	-1	0	1	1	0	0
4	0	0	0	-1	-1	0	1	-1
5	0	0	0	0	0	-1	-1	1

Figure 5.2: Vertex-edge incidence Matrix for Figure 5.1

of the entries equal to zero. Here the density is 40%, but in general, it will be $\frac{2}{|V|} \times 100$ % The main use of this matrix is that it gives constraints for linear programming formulations.

Because this is a very sparse matrix, it is likely to be wasteful of space.

One can use the matrix directly to calculate the in-degree and out-degree of any vertex.

5.3   VERTEX-VERTEX ADJACENCY MATRIX

The **vertex-vertex adjacency matrix**, or simply the **adjacency matrix**, has one row and one column for each vertex. The entry in row i and column j is one if there is an edge from i to j, and zero otherwise. If one wants to store parameters of the edges, then these can be recorded as further square matrices, usually with a zero in the matrix where there is no corresponding edge. However, it may be necessary to refer to the adjacency matrix to see whether such a zero means that an edge does not exist or whether it does with parameter value zero. Such matrices are known as **data matrices**. Figure 5.3 The advantages of this format depend on the denseness of the matrix. It is

$$
\begin{array}{c@{\;}ccccc}
 & 1 & 2 & 3 & 4 & 5 \\
1 & 0 & 1 & 1 & 0 & 0 \\
2 & 0 & 0 & 1 & 1 & 0 \\
3 & 0 & 0 & 0 & 1 & 1 \\
4 & 0 & 0 & 0 & 0 & 1 \\
5 & 0 & 0 & 0 & 1 & 0
\end{array}
$$

Figure 5.3: The vertex-vertex adjacency matrix for Figure 5.1.

$$
\begin{array}{c@{\;}ccccc}
 & 1 & 2 & 3 & 4 & 5 \\
1 & 0 & 12 & 13 & 0 & 0 \\
2 & 0 & 0 & 7 & 1 & 0 \\
3 & 0 & 0 & 0 & 20 & 5 \\
4 & 0 & 0 & 0 & 0 & 9 \\
5 & 0 & 0 & 0 & 6 & 0
\end{array}
\qquad
\begin{array}{c@{\;}ccccc}
 & 1 & 2 & 3 & 4 & 5 \\
1 & 0 & 22 & 30 & 0 & 0 \\
2 & 0 & 0 & \boxed{0} & 25 & 0 \\
3 & 0 & 0 & 0 & 20 & 50 \\
4 & 0 & 0 & 0 & 0 & 16 \\
5 & 0 & 0 & 0 & 42 & 0
\end{array}
$$

Figure 5.4: The data matrices for Figure 5.1. The boxed zero represents the value of a parameter, not the absence of an edge.

an obvious format, and so is very good for communicating details to a non-specialist, but if the matrix is sparse, then it is wasteful of storage space in computer memory. (Remember that many practical problems will be based on graphs with hundreds or thousands of vertices.) If the algorithm being used is one which examines the edges which either enter or leave a vertex, then one can perform these operations in a time which is proportional to the number of vertices (one simply has to scan through the column or row for that vertex). This may be acceptable for small, dense matrices, but it may be time-consuming when the network is sparse. Accordingly, there are two further widely-used formats for storage which are designed to help speed up the search for edges which are associated with a particular vertex.

The vertex-vertex format is not appropriate when there are multiple edges between a pair of vertices.

5.4   ADJACENCY LISTS

The **edge adjacency list** $A(i)$ for a given vertex i is the set of edges which emanate from that vertex, so:

$$A(i) = \{(i,j)|j \in V, (i,j) \in E\}$$

Similarly the **vertex adjacency list** for vertex i is the set of vertices adjacent to i,

$$\{j\}|j \in V, (i,j) \in E\}$$

The adjacency list format stores the network as a set of n linked lists, one for each vertex. The general format is to have a record (when using C++, Pascal or a similar structured language) for each edge with a pointer to further records. There is an array or ordered list of pointers, one for each vertex. These pointers point to one edge's record, and that points to further ones.

So there will be a basic pointer such a the one pictured in Figure 5.5. This can be extended to have space for several parameters, and in some cases

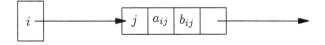

Figure 5.5: The structure of a basic pointer in an adjacency list.

the number of parameters could vary between edges. So it is a very versatile format, although it is not so obvious to a non-technical user. It can also be used to hold temporary details such as whether or not a vertex is labelled. It is obviously a good scheme for finding paths, as all the edges emanating from one vertex to adjacent ones can be found easily.

The structure for the example network will be that seen in Figure 5.6 There will normally be a list of pointers, one for each vertex, indicating the first edge in each set. Within the set of pointers whose common feature is their starting vertex, the order doesn't matter, although there will often be advantages in having some kind of arrangement. Lists and pointers provide a data structure which is adaptable. Edges can be added or removed or sets of edges can be sorted, using well-established list-handling techniques from computer science.

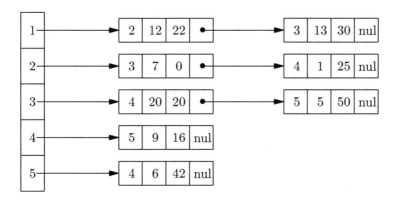

Figure 5.6: The adjacency list for Figure 5.1

5.5 FORWARD AND REVERSE STAR REPRESENTATIONS

A **forward star representation** is like the adjacency list, but it only uses two arrays. The primary (or data) array is two-dimensional, with a row for each edge and a column for each of the parameters, together with the start and finish vertices. The start vertex is normally in the first column, and the rows are arranged in ascending sequence ordered by this column. There is no need to have any kind of order on the edges whose start vertex is the same, although in most cases it will be helpful to have them in order too.

The second array is one dimensional, with one entry for each vertex. For the ith entry, this is the row number of the first edge which starts at a vertex greater than $i - 1$. (This slightly awkward definition has two consequences. First if there are no edges starting at vertex i, but there is at least one starting at $i+1$, then the ith and $i+1$th entries will be the same. Second, if the highest numbered vertex is not the start of any edges, then the corresponding entry will be the number of edges plus one.) This one-dimensional array is rather like the contents page of a book, and all the edges with the same start are identified by the first row of the primary array holding such edges, and the first row after the set which does not hold such edges. For the example network, the two arrays are shown in Figure 5.7.

The **reverse star representation** is similar, except that one will be interested in the incoming edges for each vertex, so the edges in the primary array are sorted according to their finish vertex, not their start. Figure 5.8 shows the example; again, within the basic order of edges, there is no need to sort the rows using any particular parameter or property of the edges.

$$\begin{array}{c c c c c}
 & i & j & a_{ij} & b_{ij} \\
1 & 1 & 2 & 12 & 22 \\
2 & 1 & 3 & 13 & 30 \\
3 & 2 & 4 & 1 & 25 \\
4 & 2 & 3 & 7 & 0 \\
5 & 3 & 5 & 5 & 50 \\
6 & 3 & 4 & 20 & 20 \\
7 & 4 & 5 & 9 & 16 \\
8 & 5 & 4 & 6 & 42
\end{array}$$

and

$$\text{point} \begin{array}{c c c c c}
1 & 2 & 3 & 4 & 5 \\
(1 & 3 & 5 & 7 & 8)
\end{array}$$

Figure 5.7: The forward star representation for Figure 5.1.

$$\begin{array}{c c c c c}
 & i & j & a_{ij} & b_{ij} \\
1 & 1 & 2 & 12 & 22 \\
2 & 1 & 3 & 13 & 30 \\
3 & 2 & 3 & 7 & 0 \\
4 & 2 & 4 & 1 & \cdot25 \\
5 & 3 & 4 & 20 & 20 \\
6 & 5 & 4 & 6 & 42 \\
7 & 4 & 5 & 9 & 16 \\
8 & 3 & 5 & 5 & 50
\end{array}$$

Note that there are no edges
to vertex 1 and therefore
rev point [1] = rev point [2]

and

$$\text{rev point} \begin{array}{c c c c c}
1 & 2 & 3 & 4 & 5 \\
(1 & 1 & 2 & 4 & 7)
\end{array}$$

Figure 5.8: The reverse star representation for Figure 5.1.

5.5.1 Compact star representation.

Some algorithms use the data exclusively in one or other of the star representations. However, many use data in both, sometimes using the forward representation, sometimes the reverse. It is possible, of course to maintain data in both formats during the running of an algorithm but this will involve a lot of duplication. Instead, at the cost of a little more complexity, there is a compact format which allows the advantages of both forward and reverse representations without too much repetition of the data.

In the compact format, there is an extra column in the data array which acts as an index for the reverse star. The rows of the data array are ordered in the same way as in the forward star. If they were rearranged with the index column in ascending order, then the data would fall into the same order as in the reverse star. The compact form has two one-dimensional arrays, matching the forward and reverse formats. Figure 5.9 shows the result. The star formats

$$
\begin{array}{c c}
 & \begin{array}{c c c c c}
i & j & a_{ij} & b_{ij} & \text{index}
\end{array} \\
\begin{array}{c}
1 \\ 2 \\ 3 \\ 4 \\ 5 \\ 6 \\ 7 \\ 8
\end{array} &
\left(
\begin{array}{c c c c c}
1 & 2 & 12 & 22 & 1 \\
1 & 3 & 13 & 30 & 2 \\
2 & 4 & 1 & 25 & 4 \\
2 & 3 & 7 & 0 & 3 \\
3 & 5 & 5 & 50 & 8 \\
3 & 4 & 20 & 20 & 5 \\
4 & 5 & 9 & 16 & 7 \\
5 & 4 & 6 & 42 & 6
\end{array}
\right)
\end{array}
$$

and

$$
\begin{array}{c c}
\begin{array}{c c c c c}
1 & 2 & 3 & 4 & 5
\end{array} &
\begin{array}{c c c c c}
1 & 2 & 3 & 4 & 5
\end{array} \\
\text{point} \begin{pmatrix} 1 & 3 & 5 & 7 & 8 \end{pmatrix} &
\text{rev point} \begin{pmatrix} 1 & 1 & 2 & 4 & 7 \end{pmatrix}
\end{array}
$$

Figure 5.9: The compact form for Figure 5.1; use the index column for the reverse star.

are useful in languages where there are no pointers. They are generally less efficient when it comes to changing the network, but are useful. For any of these it is possible to calculate the in-degree and out-degree values of each vertex by counting the number of entries in a column.

5.6 SUMMARY

5.6.1 Vertex-edge incidence matrix

- The storage space for (V,E) is $V||E|$

- Features:

  1. Space inefficient
  2. Expensive to manipulate
  3. Important because it represents the flow constraints on the network

5.6.2 Vertex-vertex adjacency matrix

- The storage space for (V,E) is $k|V|^2$ for some k

- Features:

  1. Suited for dense networks
  2. Easy to implement
  3. Easy to explain

5.6.3   Adjacency list

- The storage space for (V,E) is $k_1|V| + k_2|E|$ for some k_1, k_2

- Features:

    1. Space efficient

    2. Efficient to manipulate

    3. Suited for all sorts of network

5.6.4   Forward and reverse star

- The storage space for (V,E) is $k_3|N| + k_4|A|$ for some k_3, k_4

- Features:

    1. Space efficient

    2. Efficient to manipulate

    3. Suited for all sorts of network

    4. Can be used in spreadsheets

5.7   UNDIRECTED EDGES

The formats that have been described are intended for networks with directed edges. Where there are undirected edges, one can replace each by a pair of directed edges, and represent the network as one with all edges directed. This is probably the best way to handle mixed graphs. For networks where all the edges are undirected, it is often more efficient to make slight modifications to the formats. Vertex-vertex incidence matrices will be symmetric, so one only needs to store half the matrix. Each entry will correspond to one undirected edge, but any algorithms using such matrices will need to be adapted to handle the alteration. Both the adjacency list and forward star formats can be used, but it is essential to make sure that any changes to an edge are made to both the records; one way of doing this is to add extra parameters to the data which is stored in the record or line of the array, identifying the record or line for the edge in the reverse direction. Generally, the reverse star notation is not needed with networks where all the edges are undirected.

5.8   EXERCISES

1. A network has directed edges, each with two parameters, a_{ij} and b_{ij}. For each of the formats described, explain how one could find those edges for which $a_{ij} < b_{ij}$. Which format is best suited for this problem?

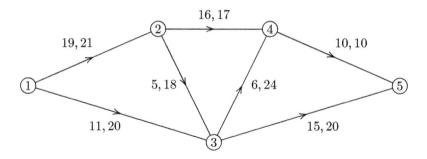

Figure 5.10: Diagram for exercise 2.

2. For the network shown in Figure 5.10, write down the vertex-edge incidence matrix and the vertex-vertex incidence matrix. Write down the representation as an adjacency list.

3. For the network shown in Figure 5.11, write down the adjacency list, forward star and compact star representations, assuming that there is a variable parameter (x_{ij}) for each edge, which is initially zero.

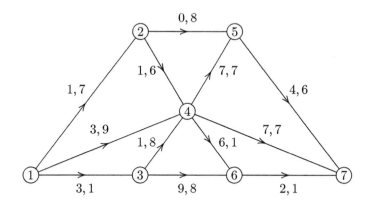

Figure 5.11: Diagram for exercise 3.

4. Figure 5.12 shows a vertex-vertex incidence matrix, with two blocks where all the entries are zero. The blocks (sub-matrices) U and W contain non-zero entries. Show that this corresponds to a bipartite graph.

5. Find a diagram of a computer network on the world-wide-web using

Figure 5.12: Diagram for exercise 4.

a search engine, and consider how the network that you find can be represented by one of the methods described in this chapter.

6. What are the advantages and disadvantages (if any) of the methods described in the chapter when the network may change by:

   • gaining extra edges with no change in the vertices;
   • gaining extra edges and vertices;
   • losing an edge;
   • gaining an extra parameter on each edge?

7. Given a network stored as a forward star, how can one decide whether or not the network possesses any isolated vertices?

8. (Computer work.) Write a computer program which will accept input of edge details i, j, a_{ij}, b_{ij} and output the forward star matrices for the network.

6

More about Shortest Paths

When you have finished this chapter, then:

- you will know about two developments of Dijkstra's algorithm, one where there are negative weights on edges, and one designed to find the shortest path more rapidly;

- you will have an introduction to the problem of finding all the shortest paths through a network.

6.1 INTRODUCTION

This chapter deals with three extensions of the shortest path concept. First, there is an algorithm for problems with edge lengths which are negative, such as may be encountered in networks representing cost or income. Income or profit from using an edge is, effectively, a negative cost. Generally, it can only be obtained once. Then, there is a short discussion about a way of speeding up the Dijkstra algorithm and finally, a presentation of ways of finding the lengths of all the shortest paths in a network.

6.2 FORD'S ALGORITHM

Dijkstra's method is a **label-setting algorithm**. Once set or made permanent, the labels can never be changed. For the examples that have been considered so far, this is not a problem. However, if there are edges with a negative cost, then fixing the labels may not result in the best path. In Figure 6.1 there is a simple example with such a profit on the edge $(2,3)$. Using Dijkstra's method, one would go directly from 1 to 3. However, the "cheapest" path from 1 to 3 is via vertex 2, since there will be a profit of £5 on edge $(2,3)$, which is large enough to offset the extra cost of going from 1 to 2 instead of directly from 1 to 3. It is worth while going off the path found by Dijkstra's method, to gain the benefit of the edge with negative weight.

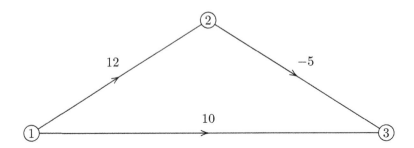

Figure 6.1: A simple network with one negative edge weight.

Ford's method is designed to solve problems with negative weights. It a **label-correcting method**. The procedure uses the same basic idea as Dijkstra's algorithm, but the labels on the vertices are never permanent; any label can, possibly, be changed at any time. It does this by a simple modification to the usual algorithm.

At any stage in the search, there will be labels on all the vertices. Then one must look through all the edges, and see if it is possible to reduce the labels on any of them. If so, then do so, and start all over again. Continue until there is no change in any label as a consequence of going through the list of edges. The algorithm starts with a label of zero on vertex s, and infinite labels on every other vertex. To protect against infinite loops, there is a rule which acts as an "emergency stop".

6.2.1 Formal statement of Ford's algorithm

0: Assign a label $l_i = \infty$ to all vertices in the network except vertex s for which $l_s = 0$.

1: Scan all the vertices k in some order, and for each one, examine all the outgoing edges, (k, j) and the labels on their destinations. For each one, calculate the smaller of l_j and $l_k + d_{kj}$ and assign this value to the label l_j. Keep a count of the number of changes that are made to labels.

2: Stop when a complete scan through the vertices and edges has failed to find any changes. The labels that are found are the minimal cost distances. *Emergency stop to check that the labels are not being changed so many times that the algorithm is going through a loop.* If step 1 has been repeated $|V|$ times, then stop, otherwise go to step 1.

Ford's algorithm automatically gives all the shortest path lengths from the starting vertex s.

6.2.2 Example

For the very simple example shown above, it is straightforward to go through the rules, as follows.

0: Assign labels

vertex	1	2	3
label	0	∞	∞

1: 1. edge (1,2), $l_2 = \min(\infty, 0 + 12) = 12$ number of changes $= 1$

2. edge (1,3), $l_3 = \min(\infty, 0 + 10) = 10$ number of changes $= 2$

3. edge (2,3), $l_3 = \min(10, 12 - 5) = 7$ number of changes $= 3$

2: There have been 3 changes, so do step 1 again.

vertex	1	2	3
label	0	12	7

1: Scan all the vertices k in some order, and for each one, examine all the outgoing edges, (k, j) and the labels on their destinations. For each one, calculate the smaller of l_j and $l_k + d_{kj}$ and assign this value to the label l_j. Keep a count of the number of changes that are made to labels.

1. edge (1,2), $l_2 = \min(12, 0 + 12) = 12$ number of changes $= 0$

2. edge (1,3), $l_3 = \min(7, 0 + 10) = 7$ number of changes $= 0$

3. edge (2,3), $l_3 = \min(7, 12 - 5) = 7$ number of changes $= 0$

2: Stop.

The test on the upper limit to the number of repetitions of step 1 can also be replaced by recording the number of times that each label is reduced, and stopping if any label is reduced $|V|$ times or more. The justification for this is that the shortest path from any vertex to any other cannot include more than $|V| - 1$ edges, so that the largest number of times any label can be reduced and still be reached by a loopless path is $|V| - 1$. So, the largest number of values possible is $|V|$. More than this, and the path must have visited some vertex twice, via a cycle of negative cost. The example in Figure 6.2 shows how finding a shortest path may require a label to be changed this maximum number of times. The complexity of Ford's algorithm is $O(|V|^3)$ because the worst case behaviour is to have to go through step 1 $|V|$ times, and the work involved in that step is similar to that for Dijkstra's method, which is $O(|V|^2)$.

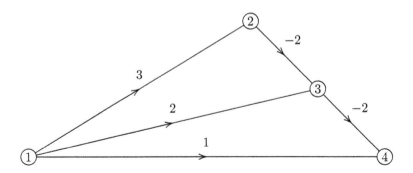

Figure 6.2: The label on vertex 4, in this network with 4 vertices, will take up to 4 values, depending on the order in which edges are examined. In the worst case, it will first be ∞, then 1, then 0, and finally -1.

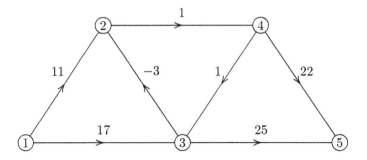

Figure 6.3: A network with a cycle of negative weight.

6.2.3   A cycle of negative weight

Figure 6.3 shows an example for which the emergency stopping rule will be needed. Here vertex 2 will be labelled 11, vertex 4 labelled 12, and soon afterwards, vertex 3 will be given a label of 13, which leads to a new label on vertex 2 of 10. This new value then leads to a succession of labels: vertex 4, labelled 11; vertex 3, labelled 12; then, vertex 2, labelled 9. And so the process will be repeated endlessly, and the label on vertex 5 will also be reduced infinitely often.

6.3   THE TWO-TREE VARIANT OF DIJKSTRA.

The speed of Dijkstra's algorithm depends on the number of vertices which have to be labelled. If one can eliminate any which are not likely to be on the shortest path, then the algorithm will run more speedily. The problem is to identify such vertices. For networks representing travel between towns and other locations, some vertices can be ignored because they are "in the wrong direction" for the path. Generally the shortest path between a point in the south and one in the north does not include an intermediate vertex which is to the south of both. The absurdity of such a route was described as part of a twentieth century poem:

> "A merry road, a mazy road, and such as we did tread
> The night we went to Birmingham by way of Beachy Head."

> (The Rolling English Road: G.K Chesterton)

The author imagined a journey from London to Birmingham, about 200 kilometres north-west, via the south coast of England, adding over 150 kilometres.

One way of identifying vertices which are not likely to be used in a path is to store their co-ordinates, and use the start and finish vertices to define a region within which all the vertices are of interest, and outside which the vertices will be ignored. This might be a rectangular box, whose corners are just beyond the start and finish vertices, or some other simple shape, such as an ellipse or circle. (Care is needed to make sure that natural obstacles such as rivers and mountains are incorporated into the design of the shape.)

It is harder to eliminate vertices from consideration when the network represents costs or time, and not distance. In many cases these measures are closely related to distances, but variations in the quality of roads, and the concentration of public transport along limited routes may mean that the route for minimal time involves a deviation of 30% or more from the shortest route in distance. For example, those living near London in the U.K. or Washington D.C. in the U.S. may often find it advantageous to go around the capital city rather than through it to reach a destination on the opposite side. Systems can be devised using multidimensional statistics to make approximate two-dimensional maps of information which can then be used to reduce the calculations needed in Dijkstra's method.

An alternative approach, not requiring any assumptions of a geographical layout, is to build up paths from the start and the destination at the same time. This creates two trees, and when they meet, then generally there will have been less computational effort than to build the tree from the start s alone. Figure 6.4 illustrates the concept behind the use of two trees. If the vertices correspond to places on a map, then the number which have a permanent label increases (roughly) as the square of the value of the label, on the assumption that the places are distributed uniformly. So there is one vertex inside the smallest circle (radius 1) centred on s in the figure, three vertices within the

next circle (radius 2), seven in the third and twelve in the fourth. These
numbers (1,3,7,12) are increasing more rapidly than a linear function, because
the area of the circles and the number of vertices inside them depend on the
square of the radius. To label t, an unknown distance D away from s, it will
be necessary to give permanent labels to about kD^2 vertices, where k is a
proportionality constant that depends on the network.

Instead, if one creates a tree of distances *from* s and *to* t, using Dijkstra's
method for each, then the two trees will "meet" about half-way between the
two vertices. Vertices there, about $\frac{D}{2}$ from each, will be permanently labelled.
This will mean permanent labels on about $k\left(\frac{D}{2}\right)^2$ for each tree, implying
$k\left(\frac{D^2}{2}\right)$ in all before there is a vertex that is permanently labelled in each
tree. Half the effort will be saved, and in some cases, the saving will be even
greater (such as where the number of vertices within a given distance of the
start or destination increases more rapidly than the square of the distance).
This algorithm was devised in the 1960's but seems to have been ignored until

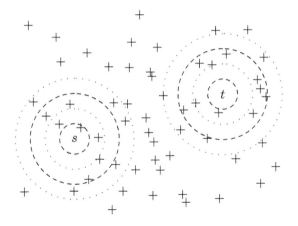

Figure 6.4: Two-trees for Dijkstra's method. The number of vertices with
permanent labels is (approximately) proportional to the square of their distance from s. Starting from both s and t halves the number of vertices given
permanent labels.

the latter part of the 1980's. It works in the obvious way, using the basic
method of Dijkstra to label vertices from s and from t. When a vertex has
been permanently labelled in both trees, then combine the results.

6.3.1 The two-tree technique

Although the steps of the algorithm are fairly obvious, the formal statement
which follows needs to include a clear test for when to stop, and how to inter-

pret the results. The stopping rule here is in step 2, when the algorithm finds if any vertices have been labelled permanently in both trees. then, in step 4, the labels are interpreted, to identify the best path, which goes from s to one such labelled vertex, and then to t; because several vertices may fall into the class, it is necessary to find the best route of all.

0: Given a network with vertices V and edges E, with directed edges with length $d_{ij} \geq 0$, define two sets of labels, ls_i, lt_i and make these all equal to ∞ except $ls_s = 0, lt_t = 0$. Define the sets $P(s), P(t)$ to be empty; these will represent the sets of permanently labelled vertices in trees rooted at s and t. Use $\overline{P(s)}, \overline{P(t)}$ to identify the temporarily labelled vertices.

1: Find the smallest valued label in $\overline{P(s)}$, m_s. Make the labels permanent on **all** the vertices with this value label, and identify the set of such vertices as $Q(s)$. Do the same for $\overline{P(t)}$, to obtain m_t and $Q(t)$.

2: Add $Q(s)$ to $P(s)$, $Q(t)$ to $P(t)$. Find if the intersection $P(s) \bigcap P(t)$ contains any vertices; if so, go to step 2.

3: For all the edges $\{(i, j) \mid i \in Q(s), j \in \overline{P(s)}\}$, let $ls_j = \min(ls_j, m_s + d_{ij})$. For all the edges $\{(i, j) \mid j \in Q(t), i \in \overline{P(t)}\}$, let $ls_i = \min(ls_i, m_t + d_{ij})$. (This is similar to the labelling process for Dijkstra's method, but is may be applied to several vertices at the same time. The star notations for storing the parameters of the networks are appropriate for this step.) Go to step 1.

4: For each vertex $i \in P(s) \bigcap P(t)$, calculate $ls_i + lt_i$ and record the smallest, and the corresponding vertex, k. Then the shortest path from s to t is the path from s to k defined by the labels ls followed by the shortest path from k to t defined by the labels lt. Stop.

Example

This algorithm will be most valuable when the network is much larger than textbook-sized examples. Any example will be artificial and should be accepted as such. Consider the network in Figure 6.5, with edge lengths shown.

(0) For initialization, the labels will be:

Vertex	s	1	2	3	4	t
ls	0	∞	∞	∞	∞	∞
lt	∞	∞	∞	∞	∞	0

and $P(s) = \emptyset, P(t) = \emptyset$

(1) $m_s = 0$ from vertex s and $m_t = 0$ from vertex t $Q(s) = \{s\}, Q(t) = \{t\}$

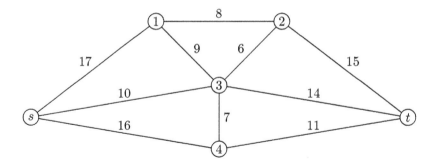

Figure 6.5: Network for the two-tree example.

(2) $P(s) = \{s\}, P(t) = \{t\}$ and their intersection is empty.

(3) The labels are changed to be:

Vertex	s	1	2	3	4	t
ls	0	17	∞	10	16	∞
lt	∞	∞	15	14	11	0

(1) Then $Q(s) = \{3\}, Q(t) = \{4\}, m_s = 10, m_t = 11$

(2) $P(s) = \{s, 3\}, P(t) = \{4, t\}$ and their intersection is empty.

(3) The labels are changed to be:

Vertex	s	1	2	3	4	t
ls	0	17	16	10	16	24
lt	27	∞	15	14	11	0

(1) Then $Q(s) = \{2, 4\}, Q(t) = \{3\}, m_s = 16, m_t = 14$

(2) $P(s) = \{s, 2, 3, 4\}, P(t) = \{3, 4, t\}$ and their intersection is $\{3, 4\}$

(4) For vertex 3, $ls_3 + lt_3 = 10 + 14 = 24$ and for vertex 4, $ls_4 + lt_4 = 16 + 11 = 27$ showing that the shortest path goes via vertex 3 and has length 24.

6.4   ALL SHORTEST-PATHS.

The algorithms which have been considered so far are all designed to find the shortest path from one vertex to another. Very often, what is needed is the matrix of all shortest distances, between every pair of vertices in the network. Quite clearly, one could look for this by repeated application of Dijkstra's method, but it is much more convenient to use matrix algebra and obtain the result directly. As a first stage, it is helpful to define a tool for collecting information from two distance matrices.

6.4.1 An operator on matrices

The new tool or operator (which is often shown by the symbol $\bigotimes$ and named "big-O-times") takes data from two $n \times n$ square matrices as follows to give a third such matrix. It is defined:

$$C = A \bigotimes B$$

$$c_{ij} = \min_k (a_{ik} + b_{kj})$$

This operator gives each element (i, j) of C the value of the smallest sum of the pairs from the ith row and jth column of A and B. The order is important, unless one is dealing with symmetric matrices. The complexity of this calculation is $O(n^3)$, as there are n^2 elements to be found, and each one requires the calculation of n sums.

6.4.2 Shimbel's method

Shimbel is credited with devising the first algorithm for finding all shortest paths in a network. It simply uses the matrix operator above repeatedly, starting with the distance matrix that corresponds to the vertex-vertex intersection matrix, D.

$$D^2 = D \bigotimes D$$

$$D^3 = D^2 \bigotimes D$$

$$\ldots \ldots$$

$$D^{n-1} = D^{n-2} \bigotimes D$$

and the matrix of shortest distances is D^{n-1}.

This works because D^2 is the matrix of shortest distances using 1 or 2 edges, D^3 correspondingly using 1, 2 or 3 edges ... and D^{n-1} using 1, 2, 3 ... n-1 edges. And no path between two vertices can use more than $n - 1$ edges. If the process is repeated any further, then the matrix D^n will be identical to D^{n-1}. The algorithm may be stopped before the maximum number of iterations if two successive matrices are identical.

In the worst case, however, the complexity of this algorithm is $O(n^4)$ which is rather excessive; the matrix operator calculates $n-2$ matrices in the sequence $D^2, D^3, \ldots, D^{n-1}$.

Figure 6.6: Shimbel's example

6.4.3   Example of Shimbel's method

Since the complexity of the algorithm is so great, any worked example will be rather tedious. Consider the very simple network shown in Figure 6.6 where the only edges are those which connect a vertex to its neighbours. The distance matrix is

$$D = \begin{pmatrix} 0 & 21 & \infty & \infty & \infty \\ 21 & 0 & 23 & \infty & \infty \\ \infty & 23 & 0 & 25 & \infty \\ \infty & \infty & 25 & 0 & 27 \\ \infty & \infty & \infty & 27 & 0 \end{pmatrix}$$

Then

$$D^2 = \begin{pmatrix} 0 & 21 & 44 & \infty & \infty \\ 21 & 0 & 23 & 48 & \infty \\ 44 & 23 & 0 & 25 & 52 \\ \infty & 48 & 25 & 0 & 27 \\ \infty & \infty & 52 & 27 & 0 \end{pmatrix} \qquad D^3 = \begin{pmatrix} 0 & 21 & 44 & 69 & \infty \\ 21 & 0 & 23 & 48 & 75 \\ 44 & 23 & 0 & 25 & 52 \\ 69 & 48 & 25 & 0 & 27 \\ \infty & 75 & 52 & 27 & 0 \end{pmatrix}$$

$$D^4 = \begin{pmatrix} 0 & 21 & 44 & 69 & 96 \\ 21 & 0 & 23 & 48 & 75 \\ 44 & 23 & 0 & 25 & 52 \\ 69 & 48 & 25 & 0 & 27 \\ 96 & 75 & 52 & 27 & 0 \end{pmatrix}$$

6.4.4   Going a little faster

There is an obvious way of speeding this up. Instead of looking at all the matrices, proceed by combining the successive ones, calculating D^2, D^4 and so on until the superscript is at least $n-1$. The number of times that matrices are combined depends on $\log_2(n)$ so that the complexity is $O(n^3 \log_2(n))$ which is better, especially for large n.

6.5   THE CASCADE METHODS

Even better, but harder to demonstrate their effectiveness, are the cascade methods. The best known one is that due to Floyd.[9]

6.5.1 Floyd's method

Instead of looking at all the options using the $\otimes$ operator and making n comparisons for each pair of vertices, Floyd proposed an algorithm which compared two distances only. The algorithm creates a succession of matrices, $L = D, L^2, L^3, \ldots, L^{n+1}$ defined as follows:

$$l_{ij}^{(r+1)} = \min\left(l_{ij}^{(r)}, l_{ir}^{(r)} + l_{rj}^{(r)}\right)$$

so that the entry in position (i, j) in the $(r + 1)$th matrix is the smaller of using the best route found so far, or using the best route from i to r and then the best route from r to j. In effect, this means that the best route found in matrix L^{r+1} goes from i to j and passes through some or all of vertices $1, 2, \ldots, r$ but not vertices $r + 1, \ldots, n$ unless the start i, the finish j, or both are greater than r.

Using the same example as for Shimbel's method,

$$L = D = \begin{pmatrix} 0 & 21 & \infty & \infty & \infty \\ 21 & 0 & 23 & \infty & \infty \\ \infty & 23 & 0 & 25 & \infty \\ \infty & \infty & 25 & 0 & 27 \\ \infty & \infty & \infty & 27 & 0 \end{pmatrix}$$

The next matrix will be identical, since there are no paths through the network via vertex 1. But L^3 will be different, because this is the matrix of shortest paths via vertex 1 or vertex 2, and (for example) $l_{13}^{(3)} = l_{12}^{(2)} + l_{23}^{(2)}$.

$$L^3 = \begin{pmatrix} 0 & 21 & 44 & \infty & \infty \\ 21 & 0 & 23 & \infty & \infty \\ 44 & 23 & 0 & 25 & \infty \\ \infty & \infty & 25 & 0 & 27 \\ \infty & \infty & \infty & 27 & 0 \end{pmatrix}$$

Continuing, the next two matrices will be:

$$L^4 = \begin{pmatrix} 0 & 21 & 44 & 69 & \infty \\ 21 & 0 & 23 & 48 & \infty \\ 44 & 23 & 0 & 25 & \infty \\ 69 & 48 & 25 & 0 & 27 \\ \infty & \infty & \infty & 27 & 0 \end{pmatrix} \qquad L^5 = \begin{pmatrix} 0 & 21 & 44 & 69 & 96 \\ 21 & 0 & 23 & 48 & 75 \\ 44 & 23 & 0 & 25 & 52 \\ 69 & 48 & 25 & 0 & 27 \\ 96 & 75 & 52 & 27 & 0 \end{pmatrix}$$

In practice, one would have to calculate L^6 as well, but from the earlier example this is known to be the optimal set of distances. Unlike Shimbel's method, one cannot stop the algorithm when successive matrices are identical. The succession of matrices depends on the manner in which the vertices have been ordered. If one has the option of arranging the vertices, then vertices of order 1 should be given the highest numbering, because these cannot be intermediate on any paths.

There is no need to have all the matrices different, and for computer implementation, it is generally sensible to use the same square matrix all the time and modify it *in situ*. However it is implemented, then this has $O(n^3)$, as may be seen in the following computer code, where the matrix L initially holds the edge lengths:

```
for k := 1 to n do
 for i := 1 to n do
 for j := 1 to n do
 L[i,j] := min(L[i,j], L[i,k]+L[k,j]);
```

(Note that the index k must be the outermost variable.)

Historically, this algorithm has a strange history. It was published very tersely in the early 1960's, simply as a piece of computer code in an American Journal which regularly printed subroutines in the language *Algol*. There was no formal proof, and it referred only to a paper on logic. As a result, it was overlooked by analysts for several years.

The idea of using one matrix only and gradually changing it led to the idea of cascade methods, such as that of Land.[15]

6.5.2   Land's method

Land's method is also $O(n^3)$; it *must* use a single matrix, which is modified as it stands. It uses the $\otimes$ operator to alter the matrix D twice. First, alter row number 1 in column order, then row 2, and so on, until the (n, n)th entry. That gives the first changed matrix.

Then work backwards, starting with the (n, n)th entry, going back to column 1 of row n, and moving up the matrix row by row, from right to left. The result of passing through the entries twice in this specific order is the desired matrix of shortest distances.

So, formally, Land's algorithm:

0: Start with the usual distance matrix, D

1: Apply the $\otimes$ operator to the first element of the first row, giving $d_{11} = \min_k (d_{1k} + d_{k1})$

2: Then $d_{12} = \min_k (d_{1k} + d_{k2})$ (noting that this involves the (possibly changed) entry d_{11})

3: Continue across the first row, then repeat with the second row, the third row and so on until the last entry in the last row. This changes D to be D^{P1} (first pass)

4: Now repeat the process, but work backwards, from entry d_{nn} to entry d_{11} to obtain a matrix D^{P2} (second pass) which is the matrix of all shortest distances.

This has complexity $O(n^3)$; it is *essential* that one works with the changed matrix, and so this is unlike the other algorithms where the result of the operator is to give the next matrix in a succession.

start	$\rightarrow$		$\rightarrow$		$\leftrightarrow$
$\hookrightarrow$	$\rightarrow$				$\leftrightarrow$
$\hookrightarrow$		$\rightarrow$			$\leftrightarrow$
$\hookrightarrow$				$\rightarrow$	$\leftrightarrow$
$\hookrightarrow$					$\leftrightarrow$
$\hookrightarrow$		$\rightarrow$			stop

First pass of Land's algorithm.

stop	$\leftarrow$		$\leftarrow$		$\hookleftarrow$
$\hookrightarrow$	$\leftarrow$				$\hookleftarrow$
$\hookrightarrow$		$\leftarrow$			$\hookleftarrow$
$\hookrightarrow$				$\leftarrow$	$\hookleftarrow$
$\hookrightarrow$	$\leftarrow$				$\hookleftarrow$
$\hookrightarrow$		$\leftarrow$		$\leftarrow$	start

Second pass of Land's algorithm.

6.5.3 Example

The example used for Shimbel's method does not demonstrate this cascade algorithm very successfully. At the end of the first pass through the matrix, Land's method has found all the shortest paths. However, if the vertices are renumbered as in Figure 6.7, then the first pass does not find every shortest path. This has distance matrix:

Figure 6.7: Land's example

$$D = \begin{pmatrix} 0 & \infty & 21 & \infty & \infty \\ \infty & 0 & \infty & \infty & 27 \\ 21 & \infty & 0 & 23 & \infty \\ \infty & \infty & 23 & 0 & 25 \\ \infty & 27 & \infty & 25 & 0 \end{pmatrix}$$

After the first pass,

$$D^{P1} = \begin{pmatrix} 0 & \infty & 21 & 44 & 69 \\ \infty & 0 & \infty & 52 & 27 \\ 21 & \infty & 0 & 23 & 48 \\ 44 & 52 & 23 & 0 & 25 \\ 69 & 27 & 48 & 25 & 0 \end{pmatrix}$$

The cascade method has not identified a path between vertex 1 and vertex 2, nor between vertex 2 and vertex 3. These are corrected in the next pass.

$$D^{P2} = \begin{pmatrix} 0 & 96 & 21 & 44 & 69 \\ 96 & 0 & 75 & 52 & 27 \\ 21 & 75 & 0 & 23 & 48 \\ 44 & 52 & 23 & 0 & 25 \\ 69 & 27 & 48 & 25 & 0 \end{pmatrix}$$

The proof is not important. The lesson from this is that one can solve the problem in $O(n^3)$ using the $\otimes$ operator. After the first pass, all the distances which use two or less edges are correct, because the method is rather like the original Shimbel method except that the changes are happening in the one matrix so the worst that might happen is that Shimbel's results will be obtained. After two passes, obviously, the matrix includes all paths with three or less edges; but the cascading actually finds all the shortest paths.

6.6   APPLICATIONS OF ALL SHORTEST PATHS

Apart from the sort of one-off process of generating mileage charts for atlases and road-books, there is a practical application in international finance of these algorithms.

Suppose that one wants to change a large sum of money into another currency. In international banking, it is possible to go directly from one to another, or via one or more intermediate currencies. The exchange rate is the product of the rates that you use (and in the world of finance, you can forget about the problems of commission faced when tourists want to change money). So the logarithm of the exchange rate is the sum of the logarithms of the rates that are used, which is just like the shortest path problem.

As the rates quoted to bankers will be varying all the time, it is necessary to have a fast algorithm for determining the best way to transfer and exchange money.

Instead of using the logarithms, you could multiply the exchange rates, not add them.

6.7   EXERCISES

1. Find the matrix of all shortest paths for the seven locations in Figure 6.8.

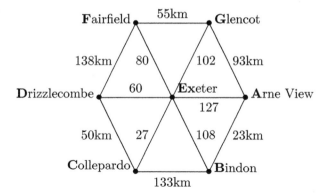

Figure 6.8: The network representing 7 locations in England's West Country, with the road distances between them.

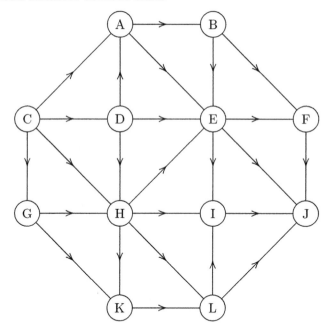

Figure 6.9: Directed network for exercises

2. As in the chapters on spanning trees and Dijkstra's method, it is relatively easy to draw a graph, pick some weights and find the shortest path as a way of gaining practice with the algorithms. Figure 6.9 gives a digraph, and Table 6.1 give the lengths of the edges, once a way of randomly assigning the letters to the numbers has been chosen. Make such a random choice, and find the shortest path from C to J.

	1	2	3	4	5	6	7	8	9	10	11	12
1	0	25	−11	9	28	5	−10	23	−29	6	32	−10
2	25	0	−5	−17	−19	−19	−3	−31	14	−17	−31	9
3	−11	−5	0	−17	5	−25	−33	−6	−22	−20	2	−27
4	9	−17	−17	0	−18	−29	−20	−24	−2	−34	−13	−4
5	28	−19	5	−18	0	−9	−1	−27	23	−13	−26	18
6	5	−19	−25	−29	−9	0	−22	−20	−7	−35	−12	−12
7	−10	−3	−33	−20	−1	−22	0	−8	−21	−23	1	−20
8	23	−31	−6	−24	−27	−20	−8	0	13	−22	−32	8
9	−29	14	−22	−2	23	−7	−21	13	0	−4	21	−21
10	6	−17	−20	−34	−13	−35	−23	−22	−4	0	−14	−6
11	32	−31	2	−13	−26	−12	1	−32	21	−14	0	16
12	−10	9	−27	−4	18	−12	−20	8	−21	−6	16	0

Table 6.1: Table of weights for exercises.

3. Consider the following proposed algorithm to find the shortest path
(s to t, as usual).

> Find the edge with the smallest cost of all, and let this cost be
> c_{min}. Replace all the costs, c_{ij}, in the network by $c_{ij} − c_{min}$.
> This will make every edge cost be zero or a positive value. Now
> apply Dijkstra's method.

Will this algorithm find the shortest path in networks with negative costs
on some edges? Explain your answer.

4. Suppose that a matrix C is formed from the vertex-vertex adjacency
matrix, giving entries of 0 for c_{ii}, 1 if the directed edge (i, j) exists, and
∞ otherwise. What is the meaning of the result when one of the all
shortest path algorithms is used with C?

5. Ford's algorithm has been used to find all the shortest paths from s
through a given network, $G = (V, E)$ with edge costs c_{ij}. How can one
find all the shortest paths when the network is modified in the following
ways? Is it necessary to start the algorithm from the beginning?

(a) An extra vertex is added, and all the edges are directed towards it;

(b) An extra vertex is added, with some edges directed towards it, and
others directed away from it; all the costs are positive on the new
edges;

    (c) An extra vertex is added, with some edges directed towards it, and others directed away from it; the costs are arbitrary, some positive, some negative, on the new edges;

    (d) The length of one edge is increased by an amount L;

    (e) The length of one edge is decreased by an amount L.

6. What happens when a network with some negative costs, but no cycles of negative cost, is used as the input for one of the all shortest path algorithms?

7. What happens when a network with a cycle of negative cost is used as the input for Shimbel's algorithm? Does the resulting matrix have any significance?

8. Suppose that the vertices of a graph represent road junctions in a regular grid network such as shown in Figure 6.10. Assume there are p roads running north-south and q running east-west, giving pq vertices. Show that the shortest path between any pair of vertices requires at most $p + q - 2$ edges and explain why the stopping rule for Shimbel's method can be changed if such a network were input. (This would not be the best way of finding all shortest paths in such a network!)

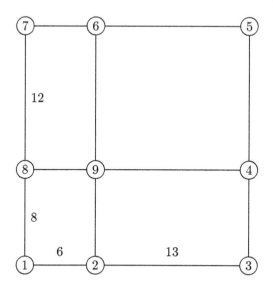

Figure 6.10: Three-by-three rectangular grid

9. With roads such as in Figure 6.10, explain why Floyd's method cannot be modified. (Hint: the labelling of the vertices in the diagram is significant.)

7

Advanced Maximal Flow.

When you have finished this chapter, then ...

- you will be able to use a modification to the Ford-Fulkerson algorithm which accelerates the search for a maximum flow;

- you will have studied a different style of maximum flow algorithm which is generally a fast way to find maximum flows;

- you will have seen the complexity of these.

7.1 INTRODUCTION

This chapter is concerned with ways of speeding up the maximal flow algorithm. When the Ford-Fulkerson algorithm was described in chapter 4 it was shown that the algorithm can be extremely slow. The reason for this was that the method chooses flow-augmenting chains in an arbitrary way. There are ways of ensuring that the algorithm converges faster than this. In this chapter, a simple modification is introduced to the labelling process which accelerates the optimization. After that, there will be an outline description of a different approach to the problem of finding the maximum flow through a capacitated network.

7.2 THE E-K MODIFICATION.

Edmonds and Karp ([8]) investigated ways of avoiding the severe computational complexity of the Ford-Fulkerson algorithm. One cause of difficulty lay in the way that a flow-augmenting chain from s to t was found. When an incomplete chain is being extended then there is a choice of vertices to be used to try and extend the chain. Any labelled vertex can be used to supply flow to an unlabelled vertex. The Edmonds and Karp (E-K) modification is a simple rule which removes that choice, by specifying the order in which the vertices are examined.

Once again, the aim is to find the largest flow that can be sent from s to t, through edges which have upper bounds u_{ij}. The flow in the corresponding edge is x_{ij} and flow is conserved at all the vertices except the start and finish. Start with all vertices unlabelled except for vertex s. Add a third part to the label, which is a sequence number showing the order in which the labels were assigned. So vertex s is given the label $(-, \infty, 0)$. Then labels are assigned to every vertex that is adjacent from vertex s, and which may be labelled, giving each of these vertices a three-part label (or "triple") which ends in a 1. The first two parts of the labels are identical to those in the Ford-Fulkerson method. Next, label every vertex that is adjacent from these vertices with a final 1, giving them triples which end in a 2. Stop, as usual, when the destination t has been labelled. The third part of the label is equal to the number of edges in the flow-augmenting chain from s, and the consequence of this method of selecting which vertex to label is that the shortest possible flow-augmenting chain will be used to make any label. In particular, the chain to t is as short as possible.

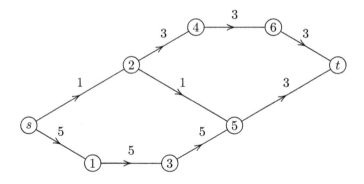

Figure 7.1: Example for the E-K modification

7.2.1   Example

The example in Figure 7.1 has been contrived to illustrate the process of labelling, but be reasonably quick. The artificial nature is needed so that there is no choice about the flow-augmenting chains at each iteration. With all flows zero, and capacities as shown in the figure, the algorithm starts searching for a flow-augmenting chain by giving s the label $(-, \infty, 0)$. The vertices that can be labelled from this are 1 and 2, with labels as shown below:

vertex	label
1	$(s, 5, 1)$
2	$(s, 1, 1)$

Then, vertices 3, 4 and 5 may be labelled:

vertex	label
3	$(1,5,2)$
4	$(2,2,2)$
5	$(2,1,2)$

From these, the last two vertices can be labelled:

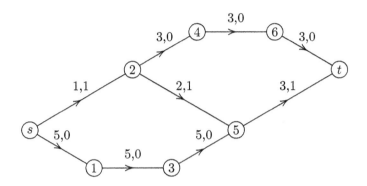

Figure 7.2: After the first flow-augmenting chain has been found

vertex	label
6	$(4,2,3)$
t	$(5,1,3)$

Hence there is a flow-augmenting chain which has a capacity of 1 unit, using 3 edges. The flow in these edges is increased, giving the flow pattern of Figure 7.2.

The algorithm removes the labels, and a new flow-augmenting chain is found. The labels will be:

vertex	label
s	$(-,\infty,0)$
1	$(s,5,1)$
3	$(1,5,2)$
5	$(3,5,3)$
2	$(5,1,4)$
t	$(5,2,4)$

Note that the edge $(2,5)$ has been used as a reverse edge to create the label on vertex 2. The second flow-augmenting chain has capacity 2 units, and uses 4 edges, demonstrating that the flow-augmenting chains are found in increasing order of length. Figure 7.3 shows the resulting flows. Once again, the labels are removed, and a fresh flow-augmenting chain is found. The labels will be:

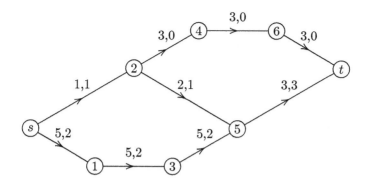

Figure 7.3: After the second flow-augmenting chain has been found

vertex	label
s	$(-, \infty, 0)$
1	$(s, 3, 1)$
3	$(1, 3, 2)$
5	$(3, 3, 3)$
2	$(5, 1, 4)$
4	$(2, 1, 5)$
6	$(4, 1, 6)$
t	$(6, 1, 7)$

indicating a flow-augmenting chain using 7 edges, with capacity 1 unit. Once the flows have been changed (including reducing the flow from 2 to 5 by 1 unit), the solution is optimal. (The cut-set is $\{s, 1, 3, 5\}$.)

7.2.2 Complexity

The complexity of the algorithm with the E-K modification is $O(|E|^2|V|)$ which does not depend on the capacity of the edges. This may be demonstrated as follows.

The capacity of a flow-augmenting chain is fixed by one of the edges on it. Call such an edge a **bottleneck edge**. It may be forward or reverse in the flow-augmenting chain. Suppose that there is a specific edge (k, l) which is a bottleneck edge as a forward edge for a chain C_1 and a reverse edge for a chain C_2. The chains will be discovered in this order, because the modification finds the short chains before longer ones. (Look at the edge $(2, 5)$ in the example to see this.)

Let $C_i(p, q)$ be the number of edges between p and q in the chain C_i; then:

$$C_1(s, l) \leq C_2(s, l)$$
$$C_1(k, t) \leq C_2(k, t)$$
$$C_1(s, t) = C_1(s, l) + C_1(k, t) - 1$$
$$\leq C_2(s, l) + C_2(k, t) - 1$$
$$= C_2(s, t) - 2$$

So $C_1(s, t) \leq C_2(s, t) - 2$, and so every time that the edge (k, l) is a bottleneck edge, then the number of edges in the chain must increase by at least 2. There are at most $|V| - 1$ edges in a flow-augmenting chain, so that the worst case is that an edge can be a bottleneck at most $\frac{|V|}{2}$ times, and hence the worst case possible is that there are $\frac{|E||V|}{2}$ flow-augmenting chains in the search.

Each flow-augmenting chain requires examination of $|E|$ edges, so the complexity of the algorithm is $O(|E|^2|V|)$ which is independent of the capacities of the edges.

7.3   PREFLOW-PUSH ALGORITHMS

There is a completely different approach to finding the maximum flow through a network. This approach has led to a family of several algorithms. All are based on the same underlying principles, so the text will only describe one. The family is referred to as being "preflow-push".

These algorithms:

- do not use flow-augmenting chains;

- create a flow out of the source which is *pushed* towards the destination;

- ignore the conservation of flows at vertices, until the end of the method;

- retain the idea of upper bounds on the edges, but that is the only constraint.

The first concept that must be defined is that of a **residual network**. When the edges in a capacitated network have been assigned flows, then there will be a new network which—effectively—shows the potential changes in flows between the vertices. So an edge (i, j) with upper bound u_{ij} and flow x_{ij} has a residual capacity of $r_{ij} = u_{ij} - x_{ij}$ and there will also be an edge (j, i) with residual capacity which can hold x_{ij} units of flow. These two numbers are those that were seen in the description of flow-augmenting chains; in some descriptions of the Ford-Fulkerson algorithm, the residual network, made up of all these effective edges, is used, and the flow-augmenting chains are found in that. The description which was used earlier is probably simpler, once the idea of a reverse edge is accepted.

Now define the **preflow** at a vertex k. This is defined for any vertex except s and t and is a set of flows in the network which satisfies

$$\sum_j x_{jk} - \sum_l x_{kl} = e_k \geq 0$$

that is the flow into k is greater than the flow out, or equal to it. If e_k is greater than zero, then k is an **active vertex**.

So if there is an edge like this, with a flow in it, then the residual network will have two edges, as shown on the right:

Figure 7.4: An edge with flow, and the corresponding edges in the residual network. For clarity, the vertices of the residual network are marked with squares.

The preflow-push algorithm creates labels which change, and which are calculated from the destination backwards. So the labels always satisfy the following rules:

1. $d_t = 0$

2. $d_k \leq d_l + 1$ for every edge (k,l) with $r_{kl} > 0$

(these labels are actually a sort of distance number of edges in the residual network.)

Edge (k,l) is **admissible** if $d_k = d_l + 1$, and the algorithm will push flows along admissible edges. (The best storage format is the forward and backward star arrangement.)

To start with, label d_i is given a value equal to the smallest number of edges that must be used in a path from i to t, except for s which is given a label equal to the number of vertices in the network. This cannot correspond to any path through the network, as was seen in the chapter on shortest paths. As the algorithm proceeds, labels on other vertices will be given similarly impossible values. The algorithm pushes flow from s to vertices closer to t. Then it repeatedly tries to push flow further on, from these vertices, towards t. When the flow into a vertex is more than can be sent further, some of the flow must be returned towards s. (It is rather as if each edge was a trough of water with an opening at the other end; splashing in one end sends a wave along the trough, and some of this goes through the opening, and some is reflected back.)

This is how it works, formally:

0: Set the distance labels to start with by searching backwards from t. Except on vertex s, the labels are as small as possible, that is they are the number of edges in the shortest path from each vertex to vertex t. So $d_t = 0$, $d_s = |V|$, $d_i = \min(d_j + 1 | (i,j) \in E)$ for each $i \neq s$. Set $x_{sj} = u_{sj}$ for all the edges going out of s. This creates one or more active vertices.

1: Select an active vertex j. If there are no active vertices, stop, with the maximum flow found.

2: If the network has an admissible edge, (j,k) then push δ units of flow from j to k where $\delta = \min(e_j, r_{jk})$.

3: If not, replace d_j by $\min(d_k + 1)$ where k is taken over the forward star of j in the residual network *and* $r_{jk} > 0$.

4: return to step 1.

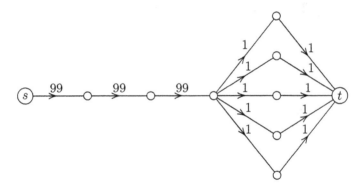

Figure 7.5: A trivial demonstration where preflow-push is useful. The edge capacities are shown.

It is possible to show that this algorithm has complexity $O(|A|^2|N|)$, but it can be speeded up by selecting the active vertex with the largest label d, in which case it has complexity $O(|N|^3)$ which is fast!

7.3.1 Why use preflows?

The example in Figure 7.5 is a trivial network where the usual maximum flow algorithm is slow. With the Ford-Fulkerson method, every flow-augmenting chain has the same three edges to start with. The preflow method changes the problem to being one where as much flow as possible has been moved to the destination.

7.3.2   Example

In the example in Figure 7.6, in which the initial flow has been forced and the excesses calculated.

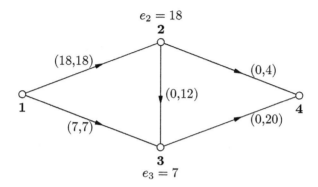

Figure 7.6: Maximum flow by preflow push; after initial forced flow. (x, u) represents the flow x and upper bound u and the two active vertices are shown with their preflows.

The residual network for this problem and these flows, together with the labels on the vertices will be as seen in Figure 7.7:

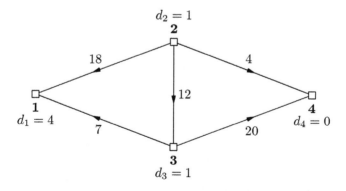

Figure 7.7: The residual network with the d labels.

Vertex 2 is active, and so the algorithm will push a flow of 4 units to vertex 4 giving the flows and excesses in figures 7.8 and 7.9:

At this stage, there is a choice, and in this example, the decision has been made to select the active vertex which has an admissible edge. So the next figures (7.10 and 7.11) are based on pushing 7 units from 3 to 4.

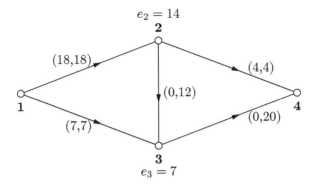

Figure 7.8: After pushing 4 units from 2 to 4.

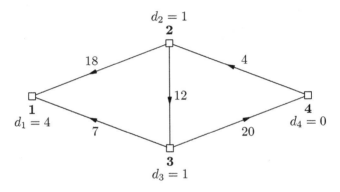

Figure 7.9: Residual flow network corresponding to Figure 7.8.

Going back to the active vertex 2, observe that it has no admissible edge, since $d_2 = 1 \neq d_3 + 1$ and $d_2 = 1 \neq d_1 + 1$. So use step 4 of the algorithm, and change d_2 to be $\min(d_3 + 1, d_1 + 1) = 2$

Then the algorithm can push 12 units of flow from 2 to 3 to give the pair of networks in figures 7.12 and 7.13:

After this, both vertex 2 and vertex 3 are active; eventually the algorithm will have to deal with both of them, and it doesn't matter which is done first. Taking vertex 2, then the only admissible edge is that which returns to vertex 1, and hence change d_2 to be 5. This is an indication that there is no longer any path from vertex 2 to the destination which can be used for sending the current excess, so the excess is returned to the source. With the new label d_2, the algorithm will push a flow of 2 units from 2 to 1, which means that vertex 2 ceases to be active and the flows and residual network are as shown

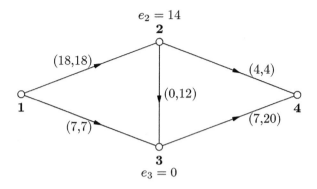

Figure 7.10: After pushing flow from 3 to 4. Note that vertex 2 is still active but has no admissible edge.

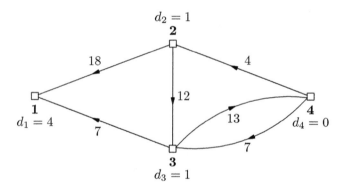

Figure 7.11: Residual network after second push, shown in Figure 7.10. Note that in the residual network, there are edges (3,4) and (4,3).

in figures 7.14 and 7.15.

Vertex 3 is active, and the algorithm finds an admissible edge and sends the excess flow from it to t, giving the situation in Figure 7.16, which is optimal, as there are no active vertices at all.

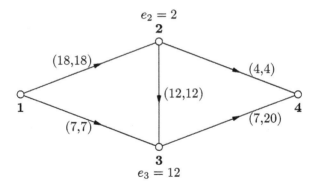

Figure 7.12: After pushing flow from 2 to 3.

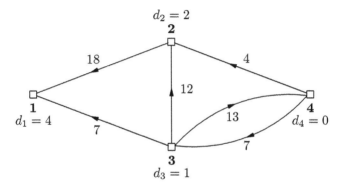

Figure 7.13: Residual network after third push. Note the changed flow has created edge (3,2).

7.4   SUMMARY AND NOTES

The preflow push method may look extremely tedious, but it is generally much more efficient than other ways of solving the maximum flow problem.

There is one step in the method where an arbitrary choice may be made, and this leads to several variations of the algorithm. The choice lies in the selection of an active vertex from which to try and push flow. Three alternative rules for this have been studied:

1. The **excess-scaling** method examines the excess at each active vertex, and finds the vertex with largest excess, and in case of ties, it selects one of those with the smallest distance label. The largest excess is e_{max}, and the active vertices are divided between those with an excess less than

$0.5e_{max}$ (the **small-excess vertices**) and the others. The algorithm tries to send flow from the vertex which has the largest excess to one of the small-excess vertices, thus ensuring a large flow is pushed towards the destination, if possible, and the large excess is reduced. In practice, further rules are applied, one of which places an upper bound on the excess which can be created at any vertex.

2. The **first-in-first-out** (or FIFO) method examines the vertices in the order in which the excesses were calculated. So the algorithm keeps a list of active vertices, selects the one at the front of the list, and pushes flow from it repeatedly, until either the vertex has no excess or the vertex has been relabelled. As flow is pushed, new vertices become active, and they are added to the end of the list. When the list is empty, then the algorithm can stop.

3. The **highest-label** method selects one of the vertices whose distance label (d_i) is greatest, and pushes flow from that. The effect of this algorithm—which is generally very efficient in practice—is to accumulate excesses and send large flows "as one batch".

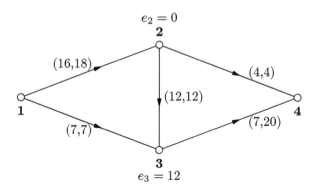

Figure 7.14: After pushing flow from 2 to 1.

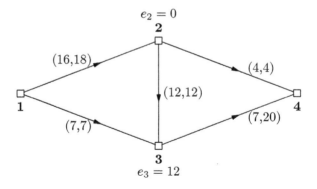

Figure 7.15: After pushing flow from 2 to 1.

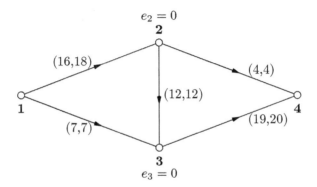

Figure 7.16: After pushing flow from 3 to 4. This is optimal since there are no active vertices.

7.5   EXERCISES

The exercises given earlier can be used again, here, and some of these exercises develop ideas which follow from the two chapters on maximum flow.

1. Find the maximum flow from s to t in the network shown in Figure 7.17 using the E-K modification to the Ford-Fulkerson algorithm.

2. Use the preflow-push method to find the maximum flow from s to t in the network shown in Figure 7.18.

3. The maximum flow through a given network is F_{max}. When an edge of the network is removed, then the flow will be F, and clearly $F \leq F_{max}$ The **least vital edge** in the network is one whose removal makes F as

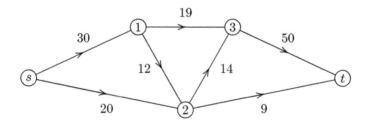

Figure 7.17:

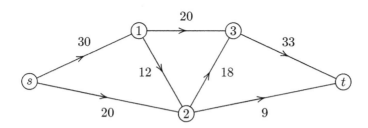

Figure 7.18:

large as possible—i.e. it has least effect on the maximum flow. Which of the following is true, and why?

(a) The edge with the smallest capacity is the least vital edge;

(b) An edge with zero flow when the flow F_{max} has been found is a least vital edge;

(c) The least vital edge cannot be in any minimal cut;

(d) The edge with the smallest flow when F_{max} has been found is the least vital edge.

4. The maximum flow through a given network is F_{max}. The capacity of every edge in the network is increased by an amount β.

(a) Prove that the maximum flow through the network is now $F_{max}+k\beta$ where k is some integer.

(b) Prove or disprove the assertion that k is equal to the number of edges in the minimal cut.

8

Minimum-Cost Feasible-Flow

When you have finished studying this chapter, you should:

- be able to describe a minimum-cost, feasible-flow problem;

- be able to describe the main steps of an algorithm which solves such problems;

- be able to recognize some of the mathematical problems for which the algorithm is suitable;

- be able to model such problems as networks.

8.1 INTRODUCTION

There are occasions when there are lower bounds on the flows in edges as well as upper bounds. An obvious example is in water supply, where the flow in rivers (and some pipelines) must be greater than some critical value or else the quality of the water will be too low. The same situation applies in cases of food mixing where it may be essential for a minimal quantity of some raw ingredient to be used in a product. It is possible to modify the maximal-flow algorithms to cope with such minima, but in general it is better to use a more powerful and more general-purpose tool. This is because there are many problems which can be modelled as networks with three parameters on each edge, the lower and upper bounds and a cost per unit of flow. This chapter is about one tool that is useful for solving such optimization problems, the out-of-kilter algorithm.

Each edge (i, j) in a network defined on $G = (V, E)$ (which may be a multigraph) has the three parameters mentioned: a lower bound, l_{ij}, an upper bound u_{ij}, and a cost per unit of flow, c_{ij}. The objective is to find flows x_{ij} which minimize the sum:

$$\sum_{\text{all } E} c_{ij} x_{ij}$$

(which is the total cost of flow in the network) while satisfying the constraints (finding a feasible flow):

$$\sum_{\text{all } i} x_{ij} = \sum_{\text{all } k} x_{jk} \quad \forall j \in V \quad \text{no special vertices}$$

$$l_{ij} \le x_{ij} \le u_{ij} \quad \forall (i,j) \in E$$

This is a linear or integer programming problem with $3|E|$ variables (including the slacks) and $|V| + 2|E|$ constraints. Because of the objective and the constraints, it is generally referred to as a "minimum-cost, feasible-flow" problem.

Although this is a linear programming problem, the fact that the constraints represent a network means that it is possible to take advantage of the structure of the network and particularly the edge and vertex constraints in order to solve the problem more efficiently. There are several ways of dealing with it. The first, and the one which is basic to many others, is the "Out-Of-Kilter-Algorithm" (OOKA) devised by Ford and Fulkerson. [10]

As was noted in the statement of the problem, there are no special sources or sinks in the network. Flow is conserved at every vertex, unlike the linear programmes that were devised for maximal flow and shortest path.

Besides the five constant parameters for each edge in such a problem (edge starting and finishing vertices, lower bound, upper bound and cost per unit) there will be one variable parameter, the flow. This means that one of the star notations will be suitable for storing the data for computation, especially as it may also be necessary to have an identifier for the edge as well, because there can be multiple edges between a pair of vertices.

As has been the case with other network problems which have an equivalent linear or integer programme, the solution method does not use the iterative approach of simplex manipulation. However, it does rely on the theory of linear programming, as will be explained later in the chapter.

When the out-of-kilter algorithm has found an optimal solution, it provides two sets of values. First are the flows, x_{ij}, one value for each edge, and second, a set of values π_i which are called the simplex multipliers and which are associated with the vertices. At the optimum, these variables satisfy one of the following three pairs of conditions on every edge:

1. $x_{ij} = l_{ij}$ and $c'_{ij} = c_{ij} + \pi_i - \pi_j > 0$

2. $l_{ij} \le x_{ij} \le u_{ij}$ and $c'_{ij} = c_{ij} + \pi_i - \pi_j = 0$

3. $x_{ij} = u_{ij}$ and $c'_{ij} = c_{ij} + \pi_i - \pi_j < 0$

These three pairs can be translated into a graphical form for each edge. In a two-dimensional space, the point $(x_{ij}, c'_{ij}$ is plotted. The conditions above represent a stepped line, which is known as the "kilter-line" for the specific edge. (Since the parameters are integers, the only values which may occur

are at integer points, but it is normal to describe the conditions as a line.)
Figure 8.1 shows such a line, for an edge whose flow must lie in the range $(3,7)$.

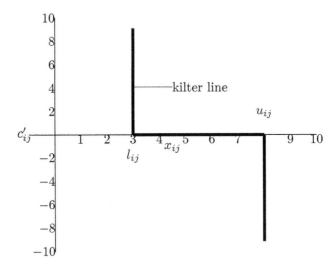

Figure 8.1: A typical kilter line.

The significance of the flows in the edges is obvious. But what about the
simplex multipliers? From a theoretical point of view, they are linked to the
dual of the linear programme. (If this introduction is baffling, then skip to the
end of this paragraph.) The structure of the linear programme described above
(the primal) means that x_{ij} occurs in four constraints, once for the flow out of
vertex i, once for the flow into vertex j, once for the constraint $x_{ij} \geq l_{ij}$ and
once for the upper bound, $x_{ij} \leq u_{ij}$. The dual problem will have a variable
for each of these four, $\pi_i, \pi_j, \alpha_{ij}, \beta_{ij}$. Given a feasible solution to the primal
linear problem, there is a corresponding solution to the dual problem. This
will be feasible so long as $c_{ij} + \pi_i - \pi_j + \alpha_{ij} + \beta_{ij} \geq 0, \alpha_{ij} \leq 0, \beta_{ij} \geq 0$. Duality
theory also imposes:

$$(c_{ij} + \pi_i - \pi_j + \alpha_{ij} + \beta_{ij})x_{ij} = 0$$
$$\alpha_{ij}(x_{ij} - l_{ij}) = 0$$
$$\beta_{ij}(u_{ij} - x_{ij}) = 0$$

The multipliers α and β can be manipulated to give the conditions for the
kilter-line above, where the reduced costs for the primal problem are the values c'_{ij}.

From a less theoretical viewpoint, the simplex multipliers serve as guides
for sensitivity analysis regarding the cost coefficient on the edges. One can
regard the value of π_i as the value of the commodity at vertex i. Then c'_{ij}

represents the cost of buying one unit of the commodity at i, moving it from vertex i to vertex j and selling it there. If this term is positive, then it will cost money to make the transfer and so one will want the amount moved to be as small as possible. If it is negative, then money is gained by making the move, and so one will make the flow in the edge as large as possible. And if this fictional cost expression is zero, then it doesn't matter whether the flow is at its lower or upper limit or anywhere in between.

Thus there is a natural interpretation of the three possible conditions which apply at the optimum. One corresponds to having the minimal permitted flow in edges, where the cost terms ensure that the flow to the destination ought to be as small as possible; one to having the maximum permitted flow in the edge, assisted by cost terms which make such flow advantageous, and the third to a flow which satisfies the bounds, assisted by a net cost of zero.

In the rest of this chapter, there will be various network problems which can be treated as out-of-kilter problems and at the way that the π values correspond to particular interpretations of the outcomes of the problems. As we do that, we'll touch on some of the ways that the out-of-kilter algorithm works, without being specific about the fine detail.

8.2 MODELLING PROBLEMS

When modelling problems as minimum-cost, feasible-flow networks, it means that there is some decision which can be represented as a variable or function of variables associated with a network. The art of modelling is to try and identify the decision and to model it as a central part of the network problem. In some cases, one must decide on parameters for the network which will make the output have the desired format.

In what follows, there will be a series of standard questions about the problem, followed by some discussion about how the decision can be mapped into a network. It is an art as well as a science. Often there will be several ways of solving the problem, and each will have advantages and disadvantages.

General questions for minimum-cost, feasible-flow problems:

- What is the decision?

- How can the decision be modelled numerically?

- What network is needed to allow this?

- What should the parameters be on the edges?

8.2.1   Shortest path.

What is the decision for the shortest path? The shortest path will be a set of edges which are used to get from s to t. The rest of the edges will not be used. So the decision is which edges are to be used.

How can the decision be modelled numerically? As has been seen earlier, one can model the problem as an integer programme with indicator variables. Each edge can be given a binary variable, which is 1 if the edge is used, 0 otherwise. A path will be a set of edges from s to t, passing through some intermediate vertices, where the path goes in and comes out.

What network is needed to allow this? Using the network for the problem, with such indicator variables, then the cost of the objective will be sum of the products of the indicator variables and the edge weights (costs or lengths). The aim is to minimize this sum, giving a problem whose objective is clearly "minimum-cost". But, the edges must form a path, so that at each vertex flow will be conserved. For each vertex there will be some flow inwards, 1 if it is used, 0 if not, and this must match the flow outwards, giving flow conservation— except at the vertices s and t. So, in order that flow is conserved at s, where there will be a natural unit flow outwards, there must be an edge with a flow of 1 unit inwards, and at t, there must be an edge to take a flow of 1 unit outwards. The same edge can be used for each, and acts as a way of returning flow from t to s. So the network for the shortest path problem needs to be extended with an edge (t, s).

What should the parameters be on the edges? There will be two types of edges in the network for the out-of-kilter algorithm. The first will be those edges which came from the original network, and in which the minimal cost flow will be found. These edges will have either zero flow, or one unit, so their lower bounds will be 0, their upper bounds 1. The second type is the extra edge, returning from t to s. This was created to take the flow back to the start, but it has a more important purpose, to ensue that there is one unit of flow going from s to t. So this edge has lower and upper bounds of one unit, forcing one unit of flow to enter vertex s, and therefore to pass through the network. The second edge is there to ensure feasibility and a relevant solution. The cost per unit doesn't matter, so it is usually set to 0.

Now the algorithm tries to find feasibility ($x_{ts} = 1$) and minimal cost (the shortest path). Conventionally, the out-of-kilter algorithm starts off with zero flows in all the edges, and zero values for all the simplex multipliers. If one does this and then draws the kilter diagrams for all the edges in this shortest path network, one would find that the edges in the body of the network were all on their kilter lines and the newly added edge was not on its.

This is illustrated in trivial case where there are three vertices 1,2 and 3; the aim is to find the shortest path from vertex 1 to vertex 3; the distances are as shown in Figure 8.2. Figure 8.3 shows the corresponding kilter diagrams with all the variables set to zero.

This yields the specification for an out-of-kilter problem with the following details:

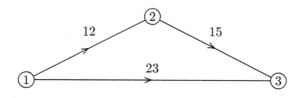

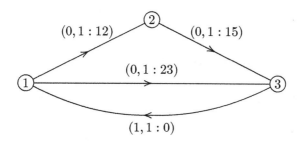

Figure 8.2: A tiny minimum-cost, feasible flow problem, based on a shortest path problem (top) and with the return edge (below).

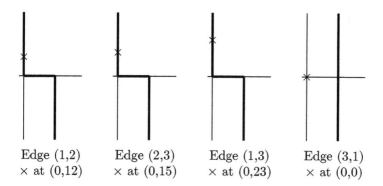

Edge (1,2) Edge (2,3) Edge (1,3) Edge (3,1)
× at (0,12) × at (0,15) × at (0,23) × at (0,0)

Figure 8.3: The kilter diagrams for the four edges in Figure 8.2, with all variables zero.

start	finish	lower	upper	cost	flow	c'
1	2	0	1	12	0	12
1	3	0	1	23	0	23
2	3	0	1	15	0	15
3	1	1	1	0	0	0

vertex(i)	1	2	3
π_i	0	0	0

(The specification of the network is being stored in a forward star matrix.)

With these initial conditions, edge $(3,1)$ is out of kilter, so it is necessary to increase the flow. To do that means sending flow from vertex 1 to vertex 3 through the network. The algorithm's rules use a labelling algorithm, trying to find a flow augmenting chain. The algorithm has the underlying principle that edges which are in kilter should never be taken off their kilter line, and edges which are out of kilter should be moved towards their kilter line, but never beyond it. To send flow out of vertex 1 means flow in one of the edges $(1,2)$ or $(1,3)$. With the present values of π, it is not worthwhile because the price of the item at vertices 2 and 3 is less than the cost of "buying" it at vertex 1 and paying for transport. The edges are on their kilter line, and a change in flow will take them off. To deal with this, the algorithm increases the price at these vertices ... and tries to do this as cheaply as possible. Changes in the simplex multipliers move the kilter positions vertically, while changes in flows move kilter positions horizontally. Here an increase in both π_2 and π_3 by 12 units will make $c'_{12} = 0$, corresponding to being indifferent about the flow in edge $(1,2)$. The kilter positions of the edge $(2,3)$ will not change. In the labelling algorithm, vertex 1 is labelled as the source of flow, and as many other vertices are labelled. The π values for unlabelled vertices are increased in such a way that at least one more vertex can be labelled, or the original out-of-kilter edge is in kilter.

Here the parameters of the problem change to:

start	finish	lower	upper	cost	flow	c'
1	2	0	1	12	0	0
1	3	0	1	23	0	11
2	3	0	1	15	0	15
3	1	1	1	0	0	+12

vertex(i)	π_i
1	0
2	12
3	12

These are pictured in Figure 8.4.

It still isn't worthwhile transporting the goods to vertex 3, so one will have to increase the price at that vertex just enough to be indifferent about transportation or not. The smallest change which could do that would be to increase π_3 by 11 and that would make $C'_{13} = 0$, allowing a flow from 1 to 3. Making this change yields values:

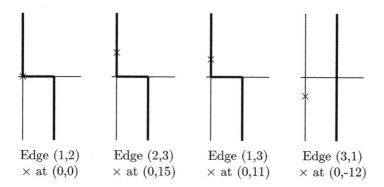

Edge (1,2)   Edge (2,3)   Edge (1,3)   Edge (3,1)
× at (0,0)   × at (0,15)   × at (0,11)   × at (0,-12)

Figure 8.4: The kilter diagrams after changing π_2 and π_3.

start	finish	lower	upper	cost	flow	c'
1	2	0	1	12	0	0
1	3	0	1	23	0	0
2	3	0	1	15	0	4
3	1	1	1	0	0	+23

vertex(i)	π_i
1	0
2	12
3	23

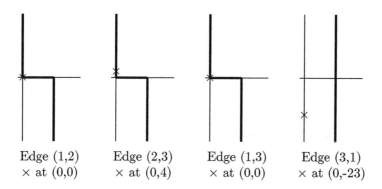

Edge (1,2)   Edge (2,3)   Edge (1,3)   Edge (3,1)
× at (0,0)   × at (0,4)   × at (0,0)   × at (0,-23)

Figure 8.5: The kilter diagrams after a further change to π_3.

These are pictured in Figure 8.5.

Now, the flows in edges $(1,3)$ and $(3,1)$ can be increased by 1 unit, making the edge $(3,1)$ in kilter, as seen in the diagrams in Figure 8.6.

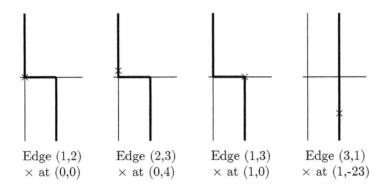

| Edge (1,2) | Edge (2,3) | Edge (1,3) | Edge (3,1) |
| × at (0,0) | × at (0,4) | × at (1,0) | × at (1,-23) |

Figure 8.6: The final kilter diagrams.

8.2.2 The out-of-kilter algorithm: formal statement

0: Assign arbitrary simplex multipliers π_i to each of the vertices i. These must be integers. Assign arbitrary flows x_{ij} to all the edges (i,j) These must be non-negative integers and the flow conservation equations must be satisfied at each vertex. It is not essential that the flow satisfies the bounds (which are $l_{ij} \leq x_{ij} \leq u_{ij}$) Often, the flows are set to zero, and the simplex multipliers to zero—but this is not essential.

1: Find an edge which is out-of-kilter, which we will identify as being the edge linking vertices s and t. If the edge is to the left of the kilter line (i.e. flow too small for the present c'_{ij}) then we identify the edge as being (t,s); otherwise it will be identified as (s,t); in each case, the edge can be made "more in-kilter" if we find a chain where flow can be increased from s to t. If there is no such edge, stop, since all edges are in kilter and an optimal flow pattern has been found.

2: Find a flow augmenting chain from s to t. The chain is found using an algorithm which is very similar to the maximum-flow algorithm. Each label is in two parts, but the way that the possible flow is calculated is different from that of the maximum-flow method. A forward edge (i,j) can be in the chain if either:

$$c'_{ij} > 0, x_{ij} < l_{ij}, \quad \text{then we set} \quad \delta_{ij} = l_{ij} - x_{ij}$$

or: $c'_{ij} \leq 0, x_{ij} < u_{ij}, \quad \text{then we set} \quad \delta_{ij} = u_{ij} - x_{ij}$

A reverse edge (i,j) can be in the chain if either:

$$c'_{ij} \geq 0, x_{ij} > l_{ij}, \quad \text{then we set} \quad \delta_{ij} = x_{ij} - l_{ij}$$

or: $c'_{ij} < 0, x_{ij} > u_{ij}, \quad \text{then we set} \quad \delta_{ij} = x_{ij} - u_{ij}$

vertex s is given a label $(-, b_s)$ where b_s is the smallest change in the flow in the out-of-kilter edge which would bring it into kilter. Then the other labels are fixed using the values of δ which is the largest potential change in the edge in the same way as in the maximum flow algorithm. If a flow augmenting chain can be found, then change the flow by the capacity of the chain, remove all the labels and return to step 1. If not, continue to step 3.

3: Examine the edges which link labelled and unlabelled vertices. For the set of edges which go from a labelled vertex to an unlabelled vertex, and for which $c'_{ij} > 0, l_{ij} \leq x_{ij} < u_{ij}$ find δ_1 as the smallest of c'_{ij}. For the set of edges which go from an unlabelled vertex to a labelled vertex, and for which $c'_{ij} < 0, l_{ij} < x_{ij} \leq u_{ij}$ find δ_2 as the smallest of $|c'_{ij}|$. If the out-of-kilter edge has either $c'_{st} > 0, x_{st} = u_{st}$ or $c'_{ts} < 0, x_{ts} = l_{ts}$ (note the directions of the edges) set $\delta_3 = |c'|$. If there are no edges in a category, then the corresponding δ-value is not set. Set δ equal to the smallest of $\delta_1, \delta_2, \delta_3$ and add δ to all the simplex multipliers on the unlabelled vertices. If the edge is still out of kilter, then go to step 2 and continue labelling. If it is in kilter, remove all the labels and go to step 1.

8.2.3 What to remember about the algorithm

Rather than try to remember the formal statement, it is sufficient to think of the following key points:

- OOKA is designed to solve the minimum-cost, feasible flow problem.

- Each edge has an integer cost, integer flow, bounded by integers.

- Flow is conserved at each vertex.

- OOKA works by:

  - Selecting x-values (flows) for each edge.
  - Selecting π-values (simplex multipliers) for each vertex.
  - Calculating $c'_{ij} = c_{ij} + \pi_i - \pi_j$ for each edge.
  - Drawing a "kilter-diagram" for each edge, plotting the "kilter-line" and the position (c'_{ij}, x_{ij}) to see if the edge is in kilter or out of kilter.

- Then, repeatedly:

  - Find an out of kilter edge and try and make it in kilter by:
  - Changing the flows in edges without making any edge out of kilter;

— If that fails, changing the π-values so that an edge becomes in kilter, or can be used to help the change of flows.

It is possible to perform simple sensitivity analysis for the out-of-kilter problem. Use the π values and the c' values which are derived from them. If one changes the cost of flow in an edge, then this will only be significant if the result is to affect the kilter diagram for that edge. So, looking at the three cases for optimality, changes in c_{ij} are only of concern when:

1: $x_{ij} = l_{ij}$ and c_{ij} is reduced so much that c'_{ij} becomes negative.

or 2: $x_{ij} = u_{ij}$ and c_{ij} is increased so much that c'_{ij} becomes positive.

or 3: $l_{ij} < x_{ij} < u_{ij}$ (**note the strict inequalities**) and c_{ij} is changed at all.

The effect of such changes is unpredictable without delving deep into the theory of the algorithm. It may mean that the simplex multipliers have to be adjusted and the flow arrangement does not change, or it might mean that the flow has to be adjusted. Going back to the shortest path problem, an example of the first of these alternatives would follow from reducing the value of c_{23} from 15 to 10. Then the values of the multipliers and the flows would have to be changed. An instance of the second case would be if any increase were to be made in c_{13}, but only if this went above 27 would there be anything more than a change in the multipliers.

8.2.4   More on shortest paths

There is a different way of setting up a shortest-path problem for the out-of-kilter algorithm. This involves the same set of central edges as before, but for the edge from vertex t to vertex s, force one unit of flow by making the cost per unit very small ($-\infty$); bounds on the flow are set to 0 and 1. The effect is the same as in the first method, although the kilter line for the return edge is stepped and not straight.

Besides being able to solve the simple shortest path problem, the out-of-kilter algorithm can be extended to a wider class of related problems. Is one wants to find the shortest paths from vertex s to several sinks: $t_1, t_2, \ldots, t_r$, then this can be done by creating a return edge for each one with parameters $l = 0, u = 1, c = -\infty$ and in the original edge, giving the upper bound on flow in every edge the value r, the number of shortest paths being looked for. In the same way, one can tackle the shortest paths from several sources to one sink. But one cannot modify the network to persuade the out-of-kilter algorithm to find the shortest paths form two (or more) distinct sources to two (or more) distinct sinks. This is because the out-of-kilter algorithm has no way of knowing which way the forced flow should go.

Nor can one use lower bounds of 1 on selected edges to find the shortest path from s to t via some intermediate edge (k, l). That appealing idea fails because the algorithm could easily find the shortest path as before with a circuit involving the intermediate edge!

8.3 MAXIMAL FLOW.

The maximal flow problem is also well suited to formulation as a minimum-cost, feasible-flow network. The same four modelling questions give ideas for transforming the problem.

What is the decision in a maximum flow problem? To find a set of flows in edges which maximise the total flow between a source s and a destination t, subject to constraints on flow in each edge, and to flow conservation at each intermediate vertex–so there is zero flow into vertex s and zero out of t.

How can the decision be modelled numerically? The flow can be given a value x_{ij} in edge (i, j), and the flow conservation follows in a straightforward way. This will give a flow with zero cost, and the problem is one of finding feasibility through the network of the maximum flow problem. Somehow, the cost needs to be incorporated to give a minimal cost problem.

What network is needed to allow this? Taking the same idea as for the shortest path, flow can be conserved at s and t by introducing a return edge (t, s). This edge plays no part in the feasibility, so long as it has a large capacity; however, it can be given a negative cost so that as flow increases, so the total cost of flow in the whole network will decrease. Therefore the search for a minimal cost solution will maximise the flow in the return edge, and hence through the whole network.

What should the parameters be on the edges? This question has, more or less, been answered in the last two paragraphs. In the original network, the edges have lower bound of zero, upper bound given by their capacity, and zero cost. In the return edge, the cost is -1, the capacity infinite and the lower bound zero.

One useful consequence of this approach is that the π-values identify the edges in the cut-set. Assuming that they are initially set to zero, the final values will be 0 or +1. π_s will be zero because that vertex is the source of all flows. For the return edge to be in kilter, π_t will have to be +1. All the other π-values are 0 or 1. For edges such as (i, j) within the network proper, then the values of π_i and π_j will be equal if it is possible to increase the flow along the edge, since that would mean indifference between buying at the price on vertex j and the price after transport from vertex i. Hence all vertices which can be labelled from s will have π-values of 0. If an edge (i, j) can not be used for further flow, there will be a price premium between the ends of the edge; one would be willing to pay more for goods at vertex j than the cost of them at vertex i plus the transport costs along (i, j) because the goods acquire a kind of scarcity value when the supply is limited. As a result, one finds that the forward edges which cross the cut go from vertices where π is 0 to vertices where π is +1. The labelled vertices will have π values of 0, the unlabelled ones will have π values of $+1$.

8.4 DEALING WITH PERSONAL DATA

The minimal-cost feasible-flow problem is sometimes used to model personal data such as might be gathered from a sample or census. When such a table is made public, it will be desirable that individuals cannot possibly be identified, so some numbers in the data will be rounded up or down, to get away from this risk. A survey might reveal that in one district of a city there is one man who is disabled, lives on his own, and collects antiques. Tables showing the people who are disabled and live on their own, and people who live on their own and collect antiques, could be used to identify that this man is disabled—which might make him a target for crime. So many census results, and surveys associated with small areas, are processed so that there are no entries of one person. In the example below, the simplest form of manipulation of data is discussed, in which every entry in a published table is a multiple of a small integer, k, usually 2 or 3. Figure 8.1 shows the data; how can the entry of value 1 be hidden, without distorting the results of the census too much? The

	Antique-collector	Non-collector	
Alone	1	5	6
Not alone	2	8	10
	3	13	16

Table 8.1: Data from a small area census.

general questions can be applied to this problem.

What is the decision? The aim is to find numbers which fit into a table, which are multiples of k, and which are as close as possible to the actual data.

How can the decision be modelled numerically? The answers will be a set of numbers, one for each cell in the table, which represent the number of people supposedly possessing the characteristics of their row and their column of the table.

What network is needed to allow this? An entry in the ith row and jth column can be modelled as a from a vertex representing all the people with the characteristics of row i to a vertex representing all those with the nature of column j. So there will be a vertex for each of the rows, connected by a set of edges to vertices for each of the columns. The flow in this edge will be the number of people in the cell I, j). At each row-vertex, there will be a source of flow, equal to the total number of people in that row, which should not be greatly different from the original total. At each column-vertex, there will be a similar demand for flow. In the same way used for the maximum flow and shortest path problems, these sources and demands can be modelled by edges, returning flow. However, in this case, the return flows must pass through a common extra vertex, so all the flow starts from one vertex, is sent to the row-vertices, through the edges representing the cells, to the column-vertices, where it is collected and sent to the extra vertex.

What should the parameters be on the edges? The manipulation of data takes place in two stages. First, the network just described is created. The data in the table will be a feasible flow in the network. Bounds are given to each edge, corresponding to the multiples of k just below (lower bound) and just above (upper bound) the current value of the cells, row totals and column totals. Then all these bounds are divided by k, giving bounds which are either identical (where the original flow was a multiple of k) or which differ by 1. There is no cost in any edge. Then a feasible flow is found, which will be an integer in each edge, and finally this is multiplied by k to give the solution.

8.4.1 Example

Taking $k = 2$ for the data in Table 8.1, Figure 8.7 shows the network.

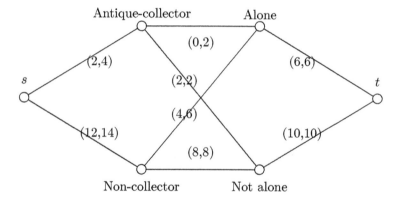

Figure 8.7:

Then Figure 8.8 shows the network where the bounds have been divided by 2. A feasible flow in this network is found, and then multiplied by 2 to give the table of less sensitive data, shown in Table 8.2.

	Antique-collector	Non-collector	
Alone	2	4	6
Not alone	2	8	10
	4	12	16

Table 8.2: A less sensitive set of results from the census; note that the row and column totals are correct, so that the manipulation of data is not apparent.

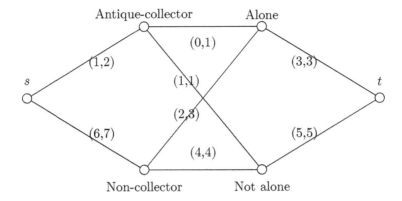

Figure 8.8:

8.5   THE TRANSPORTATION PROBLEM.

A similar network can be created for solving transportation problems as minimal-cost feasible-flow problems. (The transportation problem is to find the cheapest way to distribute some goods from factories to customers, given three sets of data: the amount that can be produced in each factory, the amount that is required by each customer, and the costs per unit carried along each permitted route.) The transport routes are edges which link vertices representing factories (sources) to those representing shops (demands).

Instead of considering all the standard questions, the description which follows concentrates on the final two, describing the network and its parameters.

Each source has an input edge, bounded above by the maximum that is available there; there is no cost on this edge. If there is a minimum amount that it is desirable to send from that source, then that should be the lower bound, but otherwise the lower bound can be zero. Each demand has an edge going out, whose lower and upper bounds should be equal to the amount required, and again no cost. These edges lead to an extra vertex where all the items are brought together, and then dispersed to the source vertices. Then the distribution edges form a large set, for each of which the lower bound on flow is 0, the upper bound is ∞ and the cost per unit is c_{ij} (where i is the source, j the demand).

With this kind of formulation, one can tackle transportation problems with extra constraints, such as bounds on the amount which goes through them.

8.6   ASSIGNMENT

A special form of transportation problem is the assignment problem, where the amounts available in each row (source) and the amounts required in the columns (destinations) are all equal to 1. The assignment problem is a highly degenerate transportation problem and it can be solved by a special algorithm (the Hungarian algorithm). However, the network formulation used for the transportation problem can be used again. (Degeneracy in $n \times m$ transportation problems means that there is a basic feasible solution which requires less than $n + m - 1$ nonzero variables. In an assignment problem of size $n \times n$, there is a basic feasible solution with n variables nonzero—considerably less than $2n - 1$.)

Assignment problems occur in a very wide range of practical applications, such as:

- Planning the make up of a swimming team for a medley relay;

- Allocating tenders between prospective subcontractors;

- Matching people to tasks or to machines or to sales areas;

- Assigning vehicles to delivery routes;

- Allocating hospital nurses to shifts;

- Matching flight crews to paired aircraft flights.

The standard problem has n people to assign to n jobs, each person i has a cost c_{ij} if they do job j. The aim is to minimize the sum of the assignment values while ensuring that every job is done and every person is employed.

8.6.1 Minimax assignment.

Because the out-of-kilter algorithm is such a rapid tool for dealing with assignment problems, it can be extended to handle a more complex variation of the standard problem. There are some occasions when the objective is not to minimize the total of the assignment values but to find an assignment whose maximum is least. It is called a **minimax assignment** or a **bottleneck assignment**. This might be the case if the cost c_{ij} is the time that person i will take to complete job j. If all the jobs are started at the same time, then the time to complete the set is the maximum of the individual times that are used. So if there were two jobs and two people, and the four possible costs were $c_{11} = 8, c12 = 6, c_{21} = 9, c_{22} = 8$, then the minimal assignment would be person 1 to job 2 and person 2 to job 1, with cost 15, and maximum 9. Exchanging the assignments would mean that the cost would become 16, but the maximum would be 8.

The problem can be solved by an iterative procedure:

1: Solve the assignment problem; if there is no solution go to step 4.

2: Find the largest cost, C say, which is used in the assignment.

3: Remove all edges whose costs are $\geq C$; this gives a new assignment problem; go to step 1.

4: The last feasible solution is the one wanted; C is the optimal value of the objective and the assignment gives the allocation of tasks.

This iterative process can be performed very quickly using the out-of-kilter algorithm. Step 3, where edges are removed, can be performed by setting the upper bound on these edges to zero, and using the previously found values x_{ij} and π_i as the initial solution, instead of zero for all variables. At least one edge will be out-of-kilter, because at least one edge with cost C was used. Hopefully, there will be relatively few such edges, and the algorithm will try to bring them into kilter, by reducing their flows from 1 to 0. The change in kilter diagrams for such an edge is shown in Figure 8.9.

Figure 8.9:

8.6.2 Assignment of groups.

A further example of a problem which is formulated as a minimum-cost, feasible-flow problem is the assignment of groups according to their preferences. In many university degree courses, students must study one or more personal projects, with an academic supervisor. The same type of problem occurs in filling sub-committees of a large committee or company board. Here, the context is of a group of S students who have to rank N supervisors or projects in some order. Each student has to take $M(< N)$ projects and there are limits on the number of students that each supervisor will look after. For simplicity, assume that each supervisor has an upper size of C_{max}, the same for each one. Assigning students to course given such data is an ideal task for the algorithm, although at first sight there isn't any apparent flow of any product anywhere. What there is a nebulous quantity which can be thought of as a student-supervisor combination. Represent each student by a vertex in a network, and each course by vertices too. The number of units of the nebulous quantity flowing through each student is exactly M, giving upper and lower bounds on a flow into each student vertex. The number of units of it flowing through each supervisor will be the number of people the supervisor can teach, bounded above by C_{max}; if there are no lower limits on supervisions, then the flow must lie between 0 and C_{max}. But of course it might be necessary to fix a practical lower limit on the supervisor. And the number of units flowing from a student to a project must be zero or one, which gives lower and upper bounds on the edges which one can use to link the people with the courses. On these edges, the cost per unit of flow could be the ranking given by the student to that course. In an ideal world, everyone would get their first M choices, which would minimize the total cost of an assignment. But this might not happen, so some people will have to take other options than these high-ranking ones.

All that remains is to have a vertex which links the source of all the units of flow *to* students to the destination of all the units of flow *from* courses. And that gives the network.

8.6.3 Classroom exercise.

A regular exercise to demonstrate the use of group assignments in the course taught by the author is to give each member of the class a list of ten potential guests for an all-expenses paid meal. Students are asked to rank these in order, giving a rank of 1 to their first choice of guest from the list, 2 to the second, and so on to the last choice. The list includes local and national figures; often the most popular are stars of TV soaps, and politicians are least popular. The number of times each guest may be selected is limited (according to the size of the class). Then the ranks are used as the costs, so without a limit on the number of meals per guest, everyone would go with their first choice guest. Using the ranks as costs is a crude way of measuring benefits, but it is simple.

The network for the group assignments is coded and solved with the out-of-kilter algorithm, and the results allow discussion of ways of making the assignment fair (each guest being "taken out for a meal" the same number of times, no student being given a guest who was ranked too low, etc.). It has been suggested that a better way is to give each student 100 units of a fictitious currency, and then they assign a sum S_1 to their favourite guest, $S_2 \leq S_1$ to the next, and so on, making sure that the total allocation is 100. Then the costs of the corresponding edges in the network will be $-S_1, -S_2, \ldots$.

8.7 KNAPSACK PROBLEMS

An integer programming problem which is of special importance in numerous circumstances is the knapsack problem. It's an all-integer problem with one constraint, and all coefficients integral. This is yet another problem which can be formulated so that the out-of-kilter algorithm is appropriate. The problem is:

Maximise:

$$Z = \sum_{i=1}^{n} C_i X_i$$

Subject to:

$$\sum_{i=1}^{n} A_i X_i \leq B$$

$$X_i \geq 0 \quad \text{and integer}$$

It helps to think about the coefficients of the objective function as being values, and the coefficients of the constraint as being units of some resource which is being used up. Suppose that there are two items, one with weight 3 kilogrammes and value 17 dollars, the other with weight 2 kilogrammes and value 12 dollars. Suppose that the limit on items is 6 kilogrammes. So, $B = 6, C_1 = 17, A_1 = 3, C_2 = 12, A_2 = 2$. Then draw a network with $B + 1$ vertices, as in Figure 8.10. In the network, there is a vertex for each possible

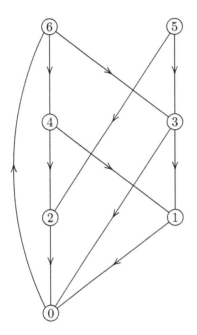

To keep the diagram free of, clutter, the parameters are not shown by the edges. On edges (6,4),(5,3),(4,2), (3,1) and (2,0) the flow bounds are 0 and 1, cost is -12. On edges (6,3),(5,2), (4,1) and (3,0) the flow bounds are 0 and 1, cost is -17. On edge (1,0) the flow bounds are 0 and 1, cost is 0. On edge (0,6) the flow bounds are 0 and ∞, cost is 0.

Figure 8.10: The knapsack problem

amount of resource, $B, B - 1, B - 2, \ldots, 3, 2, 1, 0$. For each of these, there is an edge corresponding to packing an item of type i using some of the resource. This edge goes from a vertex j to vertex $j - A_i$ and has parameters $0, 1| - C_i$. There is a return edge from vertex 0 to vertex B, and edges with no cost to vertex 0 from some or all of the vertices. (Such edges are needed from any vertex which has no other edge leaving it.) Solving the optimization problem will give flows corresponding to the items to be packed in the knapsack.

8.8   TRANSSHIPMENT.

The transshipment problem is a modification of the transportation problem to allow for some of the sources and destinations to be used as points where goods are re-routed to others. Thus the distinction between a source and a destination can become a little blurred. Obviously, this sort of problem can be easily put into the right format for the out-of-kilter algorithm and it is a good example of the general minimal cost feasible flow problem. This is a linear programme, and can be written as a transportation problem, but it is usually a hard transportation problem to solve. As a minimum-cost, feasible-flow problem, it is straightforward. Figure 8.11 shows the network of edges between factories, warehouses and shops, with the added edges for flow conservation at factories and shops, and an extra vertex to deal with all the flow. For clarity, the extra vertex is shown twice, with the dashed edge indicating the connection.

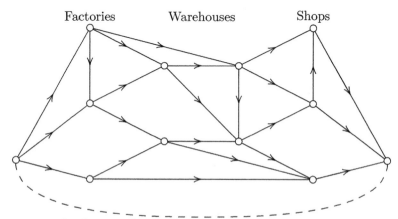

Figure 8.11: In this transhipment model, there are three factories, four warehouses and three shops. Goods may be transported between factories, and between shops. The two vertices which are connected by a dashed line are actually the same.

8.9 EXERCISES

1. Mackmurdo and Hornblower plc have leased a number of communications lines between their offices to allow data transfer between computers. An operational research study has shown that these lines can be represented as a network whose vertices (a set $N = \{1, 2, \ldots, \}$) are the offices and whose edges (a set $S = \{(i, j)|i, j \in N\}$) are the communications lines. Each edge (i, j) has a maximum capacity of u_{ij} kilobaud and costs c_{ij} per hour per kilobaud. The vast majority of the use of this network is between the two principal offices, at vertices s and t respectively. Using

the out-of-kilter algorithm, describe how you would find:

(a) the maximum communications traffic (T_{max} kilobaud) that is possible per hour between these offices;

(b) the cheapest route for sending T_{max} kilobaud between the offices;

(c) the communications line (from the set of existing ones) which would have the greatest effect on T_{max} if it had a capacity increase of 25%;

2. Why may a minimum-cost, feasible flow problem have no feasible solution? Why is it difficult to identify infeasibility?

3. A civil engineering company is building a road. Some stretches of road will run through cuttings, where rocks and soil will need to be removed; other stretches of road will run on embankments, where there will be a need for rocks and soil for filling. One 10km long piece of road has been divided into 20 stretches and the material available/needed estimated as below; the company wants to move the material at minimal cost.

Location	1	2	3	4	5	6	7	8	9	10
Available				20	66	92		38	72	65
Required	142		116				117			
Location	11	12	13	14	15	16	17	18	19	20
Available	75	96	56			67	69	66	27	
Required				126	128					180

A blank entry means that there is no material available or wanted; transport between any two locations is proportional to the number of loads and the distance between them.

Formulate the problem of moving the material as a minimum-cost-feasible-flow network problem.

4. The Laver shipping line is a company which owns a fleet of F cargo ships, all of which are to be scrapped over the next five years. If a ship is scrapped at the start of year $i = 0, 1, 2, 3, 4, 5$, the company receives income of R_i thousand pounds. In each year $i = 0, 1, 2, 3, 4$, there is a demand for D_i ships, and this is decreasing so that

$$F \geq D_0 \geq D_1 \geq D_2 \geq D_3 \geq D_4$$

The revenue from each ship that is used in year $i = 0, 1, 2, 3, 4$ is S_i thousand pounds. At the start of each of these years, the company must decide how many ships to scrap and how many ships to hire for the year, at a cost of $T_i < S_i$ thousand pounds each. At the start of year 5, the company will scrap all the ships it still owns. By appropriate definition of vertices, edges and flows, formulate the problem of deciding on an optimal (maximal profit) policy for scrapping the ships as a minimum-cost, feasible-flow network problem. Sketch an appropriate network. Explain the purpose of each vertex and edge, the constraints and the reason for these constraints.

5. Pontium Computers has six shops, selling personal computers to members of the public. It is Friday night, and in tomorrow's newspapers, there will be a special offer on the "Eclipse" system. The shops in A and B have low inventories of these, and the managers forecast that they will need 12 machines each in order to meet the expected demand. Stores in C and D have surplus machines (15 and 9 respectively). It is too late to arrange special shipping for these machines, so they will have to be sent on the regular van runs between the shops, for which there is limited space. After a short investigation of what is possible, the logistics manager has found that the following van runs could be used, and has also found the cost of transporting each system, as follows:

   (a) From D to B: capacity 6 systems, at a cost of 20 per system.

   (b) From D to C: capacity 3 systems, at a cost of 5 per system.

   (c) From C to D: capacity 7 systems, at a cost of 6 per system.

   (d) From C to E: capacity 10 systems, at a cost of 15 per system.

   (e) From D to F: capacity 12 systems, at a cost of 10 per system.

   (f) From F to E: capacity 8 systems, at a cost of 12 per system.

   (g) From F to B: capacity 8 systems, at a cost of 7 per system.

   (h) From E to A: capacity 17 systems, at a cost of 15 per system.

   (i) From B to A: capacity 7 systems, at a cost of 11 per system.

   (j) From A to B: capacity 5 systems, at a cost of 7 per system.

   The logistics manager wishes to meet the needs of the shops at minimal cost. Formulate her problem as a minimal cost, feasible flow problem, and draw the appropriate network.

6. The Ultima Thule Water Company distributes water from its storage reservoir to a water tower which supplies the city of Ultima Thule through a series of pipelines which form a network. Each pipeline (i, j) has an upper bound u_{ij}, and a lower bound l_{ij}, on the flow (x_{ij}) of water distributed per day. The lower bounds are necessary because each pipe must have a minimum flow of water as a precaution against the water becoming stagnant and polluted. The storage reservoir is identified as vertex s of the network, the water tower as vertex t.

   Describe how you would find the minimum feasible flow from s to t, by transforming the network in a suitable way so that a minimum-cost feasible-flow algorithm would yield the desired answer.

7. How may the following unbalanced transportation problem be formulated as a minimum-cost feasible-flow problem? There are m sources of goods, with capacities $b_1, \ldots, b_m$ metric tonnes and n demands for goods with requirements $c_1, \ldots, c_n$ metric tonnes and transport costs

$a_{i,j}$ pounds per metric tonne from source i to demand j. The total capacity of the sources is greater than the total demand. List the vertices
and edges, together with their parameters.

Can this formulation be extended to model (each taken separately):

(a) one transport link, from source I to demand J being forbidden
    completely;

(b) an upper bound K_I on the amount transported from one source I
    to each of the demand points;

(c) a restriction that the amount sent from one source I to demand J
    being either zero or at least some non-zero value $P_{I,J}$;

(d) a restriction that each source i should supply between $Q\%$ and $R\%$
    of its total capacity;

In each case, either explain why it is not possible to model the restrictions, or show how they may be included in the network optimization
problem. Will each extra constraint mean that there is a feasible solution
to the problem?

8. How may the knapsack problem:

$$\max C_1X_1 + C_2X_2 + C_3X_3 + C_4X_4 + C_5X_5$$

subject to

$$A_1X_1 + A_2X_2 + A_3X_3 + A_4X_4 + A_5X_5 \leq 15$$

with $A_i, C_i (i = 1, \ldots, 5)$ positive integers and $X_i (i = 1, \ldots, 5)$ nonnega-
tive decision variables, be modelled as a network optimization problem?
Can the formulation be extended to problems where there are upper
bounds on the decision variables?

9. At the University of Axminster, final year students choose a project
which they follow in the first or the second semester of the academic year.
These projects are chosen from a list produced by staff members. The
list indicates whether the project must take place in the first semester,
or the second, or either ("open timed" projects). Students apply for a
project by ranking their preferred five projects, and if necessary they
indicate a preference for the semester of any "open timed" ones. No
project can be taken by more than one student in a semester.

Staff members supervise either one or two projects each semester, and
either three or four in the academic year.

Once all the choices are known, an attempt is made to assign the projects
equitably, following the guidelines above, for the 30 students and 9 staff
in the School of Mathematics And Relevant Technology.

- Describe a minimum-cost, feasible-flow, network which could be used to assign projects to students, ignoring the constraints on staff. Sketch (in outline) the corresponding network. Identify the significance of each vertex and edge, and the reasons for selecting the parameters on the edges.

- How can this be modified so that the constraint on staff supervision "three or four in the academic year" can be included? Indicate the changes to the set of vertices and edges.

- How can the whole problem be formulated as a minimum-cost, feasible-flow, network problem? What further changes are needed to the network?

- Will these three problems ((a), (b) and (c)) always have a feasible solution? Explain your answer.

- If any of the problems ((a), (b) and (c)) has a feasible solution, will it be unique? How could you investigate alternative optima?

- How could you modify the formulation and network to allow for a member of staff taking maternity leave in the second semester, and only supervising one project in the first?

10. In a transshipment problem, material must be transported from two factories (A and B) to three depots (X,Y,Z). It is also possible to transport material from depot X to depot Y and from depot Y to depot Z. The costs of such transport per unit of material carried are as follows:

From↓ To →	X	Y	Z
A	11	16	20
B	15	10	20
X	-	4	-
Y	-	-	8

30 units are available at A and 70 at B. 30 units are required at X, 30 at Y and 40 at Z. It is desired to transport the units of goods at minimum total cost.

- Show how this problem may be formulated for the out-of-kilter algorithm by the creation of suitable edges and the identification of vertices with A,B,X,Y,Z together with the introduction of one extra vertex to ensure that flow is conserved at all vertices. For each edge, give the flow constraints and the cost coefficient; where there is a choice of values, the smallest possible should be given and the range stated.

- The out-of-kilter algorithm is used to solve this problem, starting from zero flow and zero values for all the simplex multipliers. In the optimum solution, material is sent from A to X, from B to Y and from Y to Z. Corresponding to this solution, the simplex multiplier for vertex A is 15. What are the smallest values for the simplex multipliers for the vertices B,X,Y and Z?

9

Matching and Assignment

When you have read this chapter, you should be able to:

- explain what is meant by *a matching, an assignment* and the properties of matchings;

- recognize some practical uses of matchings and assignments;

- prove the *marriage theorem*;

- outline algorithms for finding optimal matchings.

9.1 INTRODUCTION

This chapter introduces a type of graph/network problem which has many applications. Problems of matching and assignment are found in such diverse areas as finding routes for waste collection, tendering for construction projects and managing airline flight crews. Unlike many of the flow and path algorithms that have already been discussed, the way to solve matching and assignment problems depends on the structure of the graph, so there will be several methods to consider, and the choice will depend on what sort of graph is in use.

Suppose that $G = (V, E)$ is a weighted graph with n vertices. As usual, assume that the weight associated with each edge is an integer. Weights can be added together to give a total for a set of edges. A **matching** is a subset of the edges, $M \subseteq E$, such that each vertex of G is incident on at most one of the edges in M. (Note that matchings are defined in terms of the edges, although, when the matching is being shown, it is normal to show the whole graph with all the edges; those in the matching are highlighted in some way.) The **cardinality** of a matching is the number of edges in the set M. Clearly, the cardinality is a non-negative integer, and cannot be greater than $\lfloor n \div 2 \rfloor$. Figure 9.1 shows K_4 and sample matchings of cardinality 0, 1 and 2.

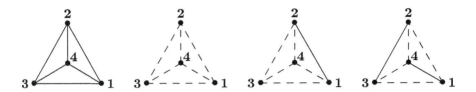

Figure 9.1: The complete graph K_4, and matchings of cardinality 0, 1 and 2. As is customary, the edges of the graph which are not part of the matching are included in the diagram; here they are shown with dashes.

When G is bipartite, a matching in G is often known as an **assignment**. Using the notation $G = (V_1, V_2, E)$, every edge in an assignment will have one end vertex in V_1 and the other in V_2. Clearly, the cardinality of such a matching will be no greater than the smaller of $|V_1|, |V_2|$.

Any matching in a weighted graph will itself have a weight, and in some circumstances this weight is more important than the cardinality. Generally, when a matching M has been found, then one can calculate the cardinality, the total **weight** and identify any **isolated vertices**, which are those without any incident edges from M. This chapter considers problems of finding the best matching, where "best" may mean maximal cardinality, maximum weight or minimum weight for a given cardinality. Maximising the cardinality also minimises the number of isolated vertices, since the number of vertices which are isolated is the difference between the number of vertices and twice the cardinality.

9.2   APPLICATIONS

Many of the applications of matching are found in trying to assign the members of one group of items to suitable members of another. What goes into those groups can be extremely varied. In each case, an item from the first group may be compatible with some of the items in the second, and there may be a cost or other weight associated with the link between the items. The examples below are typical situations; these often occur as part of much larger commercial and practical problems, and the last examples are ones which shows the financial benefits of being able to solve matching problems.

- The two groups may be people; in universities where it is the norm for entering undergraduates to share rooms, it is desirable to try and match those students to try and find compatible pairings.

- One group may be people, and the second, pieces of work. A manager may wish to assign employees to the work, making sure that all the work is done, and that suitable employees are allocated.

- Electronic equipment is often used in matched pairs (such as in stereo amplifiers and speakers). The manufacturing process has some inherent statistical variation, and so it is useful to try to match items which are as similar as possible.

- Several pieces of work can be split into two separate parts, such as a morning shift and one in the afternoon. Matching these to people, in such a way that the workload is fairly balanced is another application of matching algorithms.

- There are many types of product whose value (or quality) decreases with age. If you have a stock of such items, and a schedule for using them, then the problem of deciding when to use each item so as to use the set with the greatest value can be written as an assignment problem, matching items to the times in the schedule. Foodstuffs, and items which evaporate, are examples.

- Many pieces of scanning medical equipment try to match observations on a patient with a library of known observations, so as to help in diagnosis. Here the groups will be observations which are unknown and those which are known.

- The major airlines must match aircraft to flight routes, so that the workload on employees is acceptable, there are crews (and aircraft) for all the routes, and the cost of the allocation is optimal. With a typical airline running hundreds of routes, and employing thousands of staff, the potential for saving money by finding the best match of people and routes is enormous. (In the days after the terrorist attack on New York in September 2001, all commercial flights in the United States were grounded. The airlines used matching and assignment algorithms to restore their routes afterward.)

9.3 MAXIMUM CARDINALITY

9.3.1 Bipartite graphs

Finding a matching of maximum cardinality in a bipartite graph, $G = (V_1, V_2, E)$, is straightforward. All the edges in matchings in such graphs can be thought of as flowing from V_1 to V_2, and therefore the problem is to find a way of sending the largest possible flow from V_1 to V_2. The problem becomes one of identifying the maximum possible flow through a network with multiple sources (V_1) and multiple sinks (V_2); any maximum-flow algorithm can be used, with the modifications discussed earlier in chapter 4. (Since the edges in a matching problem are undirected, then one may interchange the two sets of vertices, and obtain the same result.) For simplicity, assume that the selected method is a labelling algorithm such as that of Ford and Fulkerson. ([10])

The steps one must take are therefore:

1. Direct all the edges to run from V_1 to V_2

2. Give each edge a capacity of infinite flow

3. Create a supersource s and a supersink t and add edges of capacity 1 unit from s to every vertex in V_1 and from every vertex in V_2 to t

4. Solve the problem of finding the maximum flow from s to t

When the maximum flow has been found, its value will be the maximum cardinality for the graph G. The edges linking V_1 and V_2 which have a unit flow will identify the matching. This is illustrated in figures 9.2 and 9.3.

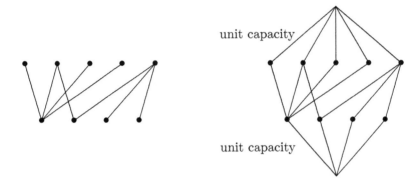

Figure 9.2: A bipartite graph, and the edges which are added to allow the maximum cardinality matching to be found using a maximum-flow algorithm.

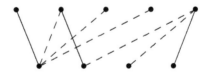

Figure 9.3: One of the maximum cardinality matchings in the bipartite graph of Figure 9.2.

9.3.2   Hall's Theorem (The "Marriage Theorem")

For a complete bipartite graph, even this simple process of finding the maximum flow is unnecessary. It is always possible to find a matching in a complete bipartite graph with cardinality equal to min $(|V_1|, |V_2|)$, taking an edge which links one vertex from V_1 to one from V_2, discarding these vertices and repeating until there are no vertices in one of the sets.

What happens if the bipartite graph G is not complete? Under what circumstances can one find a matching whose cardinality is as large as possible? Such a matching is known as **complete**. A theorem proved in 1935 by Hall[11] is useful. Assume that V_1 is the smaller of the two vertex sets (or choose arbitrarily in the case of equal sets). Suppose that A is a subset of V_1 and define $\Gamma(A)$ to be the set of vertices in V_2 which are connected to A by an edge of G. Then:

Theorem 4 Hall's theorem. *A bipartite graph G has a complete matching if and only if, for every $A \subset V_1, |\Gamma(A)| \geq |A|$.*

Proof. If there is a complete matching, then every vertex in V_1 will be matched to a vertex in V_2 and so every subset A of V_1 will be matched to a subset of V_2 of the same size. $\Gamma(A)$ will be at least as large as this. So $|\Gamma(A)| \geq |A|$ for every subset A for complete matchings.

On the other hand, if the matching M in G found by the maximum flow algorithm is not a complete matching, so $|M| < |V_1|$ consider the set B of labelled vertices. It is clear that s will be in B, and t will not. Define the set $A = V_1 \cap B$, the labelled vertices in V_1. All the edges from A towards V_2 are of infinite capacity, so that vertices in $\Gamma(A)$ are labelled, i.e. $\Gamma(A) \subset B$. All the edges from s to $V_1 \setminus A$ will be full, and so will the edges from $\Gamma(A)$ to t. But, no vertex in the set $V_2 \setminus \Gamma(A)$ can be labelled, because there is no edge connecting it to a labelled vertex. Then the cut defined by the set B and its complement $\overline{B}$ is the union of two disjoint sets of edges, those linking s to $V_1 - A$ and those linking $\Gamma(A)$ to t. The capacity of the cut is equal to $|M|$ and is the sum of the capacities of the two sets of edges, so:

$$|M| = |V_1 - A| + |\Gamma(A)|$$
$$= |V_1| - |A| + |\Gamma(A)|$$

But : $\qquad |M| < |V_1|$

so : $\qquad |\Gamma(A)| - |A| < 0 \qquad \rightarrow \qquad |\Gamma(A)| < |A|$

Q.E.D.

Figure 9.4 shows the cut.

This theorem is often called the **marriage theorem**. The name is taken from the following scenario. One set of vertices in the bipartite map represents a group of people of the same gender, and the other set represents people of the other gender. An edge links those people who are mutually attracted, and are, presumably, prepared to marry each other. Hall's theorem says that it will be possible to match couples provided that no individual or group is too limited in their choice.

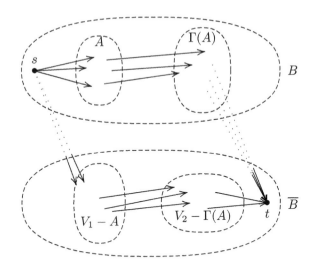

Figure 9.4: Matching in a bipartite graph. The cut when the maximal flow algorithm terminates. Edges which are dotted cannot be used to increase flow because they are full with one unit of flow. There may be edges from $V_1 \setminus A$ to $\Gamma(A)$ but these have zero flow and cannot be used to label any other vertices.

9.3.3 Example

The diagram in Figure 9.5 shows this matching problem in a small bipartite graph. The graph has vertices $V_1 = 1, 2, 3, 4, 5$ and $V_2 = 11, 12, 13, 14, 15, 16$. Edges between these are shown and the additional vertices s and t with their associated edges have been added. There are several matchings of cardinality 4 in the graph, and the diagram assumes that the matching using the edges marked "M" has been found. Vertex s is labelled as the source of flow. From this, vertex 2 can be labelled since no matched edge is incident on it. Vertex 11 can be labelled, using the forward edge (2,11). The last vertex to be labelled, vertex 1, is labelled from 11 using edge (1,11) as a reverse edge. In this example, therefore, $B = \{s, 1, 2, 11\}$, $A = \{1, 2\}$ and $\Gamma(A) = \{11\}$. $|A| = 2, |\Gamma(A)| = 1$. The edges which cross the cut $B, \overline{B}$ are all carrying one unit of flow.

9.3.4 A special case of Hall's theorem

Hall's theorem allows one to find whether or not there is a complete matching without using the maximum flow algorithm. However, it is a cumbersome way of doing so, because one may need to examine every possible subset A of V_1. The following is a useful special case.

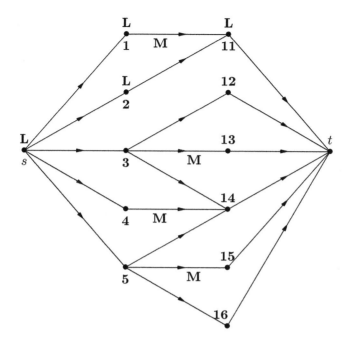

Figure 9.5: Example of Hall's theorem. Edges marked "M" are matched, vertices marked "L" are labelled from s.

Theorem 5 *Suppose that in the bipartite graph $G = (V_1, V_2, E)$, every vertex in V_1 is adjacent to at least k vertices in V_2, and that no vertex in V_2 is adjacent to more than k vertices in V_1. Then there will be a complete matching in G.*

Proof: Take any $A \subset V_1$. There will be at least $k|A|$ edges from A to $\Gamma(A)$ but there will be no more than $k|\Gamma(A)|$ edges from $\Gamma(A)$ to V_1 some of which will be incident on vertices in A. This second set of edges includes the first set, so:

$$k|A| \le k|\Gamma(A)| \quad \Rightarrow \quad |A| \le |\Gamma(A)|$$

This is true for all subsets A, and so Hall's theorem applies. **Q.E.D.**

Hence, in the scenario giving rise to the nick-name of "Marriage theorem", if every male likes at least three of the females, and the females are more selective, and none like more than three males, there will be a complete matching.

Example

Seven lecturers (Alan, Beth, Chris, Dave, Ed, Fiona, Guy) share the teaching of five university lecture courses (MAS101, MAS102, MAS103, MAS104, MAS105). The allocation is in Figure 9.6. One lecturer must be appointed

MAS101	MAS102	MAS103	MAS104	MAS105
Alan	Alan	Ed	Dave	Beth
Beth	Ed	Dave	Chris	Chris
Guy	Guy		Fiona	Fiona

Figure 9.6: Lecturers and courses; who sets the examinations?

as the examiner for each course; the lecturer must have taught on the course, and no lecturer must set more than one examination. Can it be done?

Representing the information as a bipartite graph, with vertices for each course and for each lecturer, and edges for those who teach on the courses, one obtains the graph in Figure 9.7 The theorem just seen shows that there will be a complete matching using $k = 2$. Each course is taught by at least two people, and no person teaches on more than two courses.

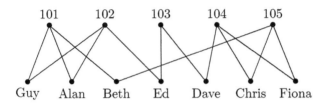

Figure 9.7: Bipartite graph of lecturer/course links.

9.4   GENERAL GRAPHS AND EDMONDS' ALGORITHM

With a general graph, there is no simple way to convert the matching problem into a flow problem. Although the objective function in a matching problem is a linear one, and many of the constraints are linear, linear programming is not a sensible way of solving general matching problems. To illustrate the difficulty, one may consider a very simple graph, such as that in Figure 9.8.

The natural way to look for the best matching with a linear programme would be to define $x_{ij} = 1$ if edge (i, j) was in a matching, and $x_{ij} = 0$ if not. The objective would be to make the sum

$$Z = \sum_{\text{edges}} x_{ij}$$

as large as possible. Clearly, there will be a constraint at every vertex i to make sure that the number of matched edges incident on i is no more than 1. This linear constraint would take the form:

$$\sum_{j} x_{ij} + \sum_{k} x_{ki} \leq 1$$

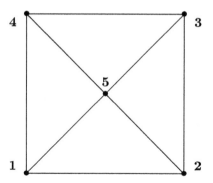

Figure 9.8: A linear programme for matching in this graph would find one of cardinality 2.5, which would not be feasible.

But in Figure 9.8, the linear programme's solution would give $Z = 2.5$, by making five of the x-values equal to 0.5. (There are several such solutions. One is: $x_{12} = x_{23} = x_{34} = x_{45} = x_{51} = 0.5$) Integer programming would give a matching with $Z = 2$; in general, the use of integer programming will not be advisable, because many graphs will rapidly create a very large branch-and-bound tree for moderate sized problems.

Because linear programming is not an advisable approach, this section presents Edmonds' algorithm, published in 1965.[7]

First, some definitions. Suppose that M is a matching in a graph $G = (V, E)$. A vertex in V will either be **matched** or **exposed** in M. A matched vertex has one incident edge, an exposed vertex has none. Choosing an exposed vertex, one can find a simple path (which may be only one edge long) which alternately uses edges that are out of the matching and in the matching. A simple path like this is called an **alternating path**. An alternating path that starts and finishes at exposed vertices is called an **augmenting path**. (Exposed vertices are also known as **free vertices**.)

With an augmenting path, there will be an odd number of edges, with (say) r edges in the matching and $r + 1$ which are not. Exchanging the status of these will give a matching whose cardinality is one greater than that of M. (The exposed vertices will become matched, and the matched vertices will remain so.) Hence, if a matching possesses an augmenting path, then it cannot be a maximum-cardinality matching. As the proof below shows, the converse is also true, which leads to Edmonds' algorithm.

> **Theorem:** A matching M is a maximum-cardinality matching, if and only if it does not possess an augmenting path

Proof: (The result above shows that a matching with an augmenting path cannot be maximal. So what follows proves that if a matching M is not maximal, then there is an augmenting path in M.) As usual, $G = (V, E)$ is the underlying graph. M is a matching for this graph, and one has chosen another matching, M^*, of maximum cardinality, and this has been chosen to have the largest number of edges in common with M of all such maximum-cardinality matchings. (How one has found M^* doesn't matter; it is enough to have defined such a matching.)

Consider the edges which are different, and create the subgraph which includes all of these. Call it G'. ($G' = (V, E')$ where $E' = \{e \,|\, e \in (M \cap (E \setminus M^*)) \cup (M^* \cap (E \setminus M))\}$.)[1]

In this subgraph, there will be at most two edges incident on any vertex; this follows from the definition of a matching, since in neither M nor M^* can there be more than one edge incident on a vertex. Therefore, the components of G' must be one of:

- Isolated vertices (degree 0);

- Simple paths, connecting an initial and terminal vertex each of degree 1 with vertices of degree 2 in between;

- Simple cycles, where all the vertices are of degree 2.

Isolated vertices play no part in the proof. The next step shows that simple cycles cannot exist in G'. Suppose that there is an even cycle. The edges in this will alternately belong to M and M^*. However, if the edges from M^* were replaced by edges from M, then the result would still be a matching, with the same cardinality, but with more edges in common with M than were possessed by M^*. However, M^* was chosen to have as many edges in common with M as possible, so there is a contradiction and so an even cycle cannot exist. On the other hand, the cycle cannot be an odd one. If it were, then one vertex would be incident on two edges from the same matching, which contradicts the definition of matching.

So the only components of interest in G' are simple paths. If one finds a path with an even number of edges, then the same argument as for simple cycles applies; the edges alternate from the two matchings, and one could find a matching with more edges in common by an exchange of edges. So the paths must be odd paths, and the first edge and last edge come from the same matching. This matching cannot be M, otherwise the path would be an augmenting path for the maximum-cardinality matching M^*. So the first and last edges in this odd path come from M^*, meaning it is a flow-augmenting path for M. **Q.E.D.**

Figure 9.9 shows a matching which is not maximal, one which is maximal with as many edges in common as possible, and the subgraph G' found in this theorem.

---

[1] This is one of those rare occasions when an explanation in words is shorter than reading the mathematical notation.

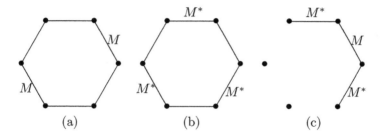

Figure 9.9: To illustrate Edmond's theorem; in (a), the matching M has cardinality 2; the matching M^* in (b) has maximum cardinality 3. The graph G' in (c) has isolated vertices and an augmenting path.

9.4.1 The tree-growing algorithm

A consequence of the previous discussion is that one can find a maximum cardinality matching by repeatedly searching for augmenting paths from any initial matching. When it is impossible to find an augmenting path, then the problem will have been solved. An augmenting path requires an exposed vertex as the start and finish. The algorithm systematically examines the exposed vertices, and tries to grow an augmenting path from each. When one is found, the matching is augmented, the vertices are no longer exposed, and the process continues. The process of growing a path actually involves growing several paths (just as in Dijkstra's shortest path algorithm) in a tree. The tree will be an **augmenting tree** rooted at the selected exposed vertex. Vertices in the tree can be classified as **outer** or **inner**, depending on whether the last edge in the path from the root is in the matching or not. The root is an outer vertex, so any path in the tree passes from outer to inner vertex and inner to outer vertex alternately.

It is convenient to consider the algorithm for growing trees separately as it will form a crucial component of the matching algorithm.

Tree-growing algorithm (from a specific root)

0: The input for the algorithm is a matching M and an exposed vertex v and an empty tree T. Define v to be the root, add it to T, and label it as an outer vertex of the tree. All the other vertices are **unlabelled**, and all the edges are **unmarked**.

1: Examine each of the outer vertices, to see if there are any unmarked edges incident on them. If an examination of all the outer vertices shows that none is incident on an unmarked edge, go to step 3, with a **Hungarian tree**. If there is one, choose it, say (i, j), with an outer vertex i, and consider vertex j. This can be one of four types, inner (and labelled), outer (and labelled), unlabelled-and-exposed or unlabelled-and-matched.

If j is inner, then mark the edge as being not in the tree, and go back to the start of this step, looking for another unmarked edge.

If j is outer, then mark the edge as being in the tree T, and go to step 2 with an odd cycle.

If j is unlabelled-and-exposed, then mark the edge as being in T, and stop with an augmenting path has been found from v to j

If j is unlabelled-and-matched, then mark the edge as being in T, find the edge (j, k) which is in M, mark it as being in T, and label j as an inner vertex and k as an outer vertex. Go back to the start of this step.

2: (Odd cycle.) Stop because an edge has been marked in the tree with two outer endpoints. This has created a cycle of edges which are in the tree, because the procedure stopped with a vertex which had already been labelled, and it must be an odd cycle, since the vertices have been alternately inner and outer until this edge was found.

3: (Hungarian tree.) Stop because no further marking is possible. The edges form a tree that is called a **Hungarian tree**.

This algorithm stops with one of three results; an augmenting tree, a Hungarian tree, or an odd cycle. If it stops with a Hungarian tree, then there can be no augmenting tree from the selected root.

9.4.2   The matching algorithm for maximum-cardinality

This tree-growing algorithm is a building-block in Edmonds' method. in the method, odd cycles are referred to as **blossoms** and they may be contracted to a single **pseudo-vertex** and then later expanded. Figure 9.10 shows a cycle with three edges, which can be reduced, and the pseudo-vertex will be labelled outer.

This algorithm starts with an arbitrary matching, M, which may consist of no edges. However, in practice, it is sensible to try and find as large a matching as possible by any convenient technique, as this saves effort later on. (A heuristic method can be used; order the vertices in increasing degree, and work through them in this sequence. If there is a vertex which is not matched, then try to find an edge which could be inserted into the matching, choosing the edge which leads to the lowest degree unmatched vertex available.) Then repeatedly the algorithm takes exposed vertices, and uses the tree-growing procedure. When augmenting trees are found, then the matching is changed for one of higher cardinality. When odd cycles are found, the graph will be shrunk,

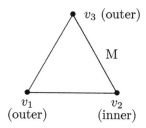

Figure 9.10: An odd cycle with three edges, which may be reduced to a pseudo-vertex

replacing all the edges and vertices of the odd cycle by one pseudo-vertex (labelled outer), and the search for augmenting paths will be resumed. At the end, the graph will be expanded back to its original size. When Hungarian trees are found, then their edges will not be considered again for the given root, which is equivalent to removing them temporarily from the graph G within the algorithm.

0: The input is a graph G and a matching M, which may be of zero cardinality. Create a graph (which acts as a stack in computational terms) K to store subgraphs of G, and make it empty.

1: Identify all the exposed vertices of M, say $v_1, v_2, \ldots$

2: Take one of the exposed vertices, v_i, put this into an empty tree T and label it outer. Clear the stack and make every edge of G unmarked.

3: Step T: Use the tree-growing algorithm to try and expand T.

4: If a blossom is found, then:

4.1: place it in K;

4.2: shrink it to an outer pseudo-vertex in G;

4.3: if the blossom includes v_i, then label the pseudo-vertex exposed;

4.4: go to step T;

5: If a Hungarian tree is found, then place it into the stack K; this has dealt with the exposed vertex v_i;

6: If there are any more exposed vertices, return to step 2; otherwise, stop with the matching M as the answer.

7: (This step is only reached if an augmenting path has been found.) Identify the path in T, and empty K item by item. Blossoms are expanded, and the path is extended by the insertion of the appropriate even-length section of each blossom. Increase the cardinality of M by interchanging matched and unmatched edges of the path. Go back to step 1.

9.4.3 Complexity

In Edmonds' paper, he wrote that the algorithm had complexity $O(|V|^4)$; later work on this algorithm has found a way of reducing this to $O(|V|^3)$. The proof is outside the scope of this chapter.

9.4.4 Example

Figures 9.11–9.14 show the progress of Edmonds' algorithm applied to a simple graph.

9.5 MATCHINGS OF OPTIMAL WEIGHT

The algorithms that have been described are all concerned with the maximum-cardinality matching problem. For some of the applications, the edges of the graph are weighted, and the objective is to find the matching whose total weight (measured by summing the weights of the edges) is maximum or minimum. Usually, the matching must be complete, or have a specified cardinality. (A perfect matching is one where the cardinality is $|V|/2$.) The examples of assigning people to jobs, and airline crew to routes are typical. Any assignment has a value, and one's objective is to find the best.

Once again, the solution method depends on whether or not the graph is bipartite. If one has a bipartite graph, $G = (V_1, V_2, E)$, with $|V_1| \leq |V_2|$, and the objective is to find a complete matching of minimal weight, then the methods described earlier for the assignment problem apply. Each edge in the bipartite graph is treated as a capacitated edge, with lower bound of 0, upper bound of 1, and the cost given. These will be directed from V_1 to V_2. The graph is extended to form a network suitable for a minimal-cost, feasible-flow algorithm, such as the out-of-kilter algorithm. There will be an extra vertex, s, acting as both a super-source and a super-sink. There will be edges from s to each vertex in V_1, with a forced flow of 1 (lower bound equal to upper bound equal to 1) and cost of zero, and edges from each vertex in V_2 to s, with zero cost, upper bound of 1 and lower bound of 1. Then, the cost will be the sum of the weights of the edges which are used, and the flow constraints mean that each vertex in V_1 is matched to one vertex in V_2. (Note that this is slightly different from the discussion in Chapter 8 using the out-of-kilter method, since here there is the possibility that the two parts of the bipartite graph have different numbers of vertices. As before, the Hungarian algorithm of König can also be used.)

9.5.1 Example

A car rental company has six cars available for customers in the coming week, and four customers who have arranged to rent a car. Because of the times when the customers made their reservations, the expected profit from the

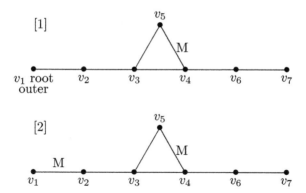

Figure 9.11: Example of matching: [1] shows an initial matching of cardinality 1; taking v_1 as root leads to an augmenting path (consisting of one edge) leading to the matching in [2]

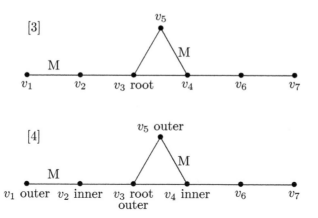

Figure 9.12: Taking v_3 as the root in [3], labelling v_4 then v_5 reveals an odd cycle in [4].

rentals vary. Table 9.1 shows which customers are willing to take which cars, and what the expected profit from each allocation will be. Which allocation makes the greatest expected profit? The solution to this network problem is to assign Meg to car R, Neil to car T, Owen to car W and Paul to car U. The profit will be 42 units. As may be expected, most of these assignments are to the most profitable rental.

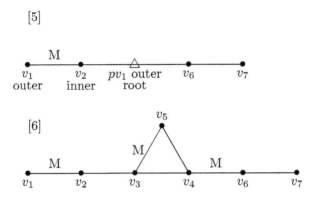

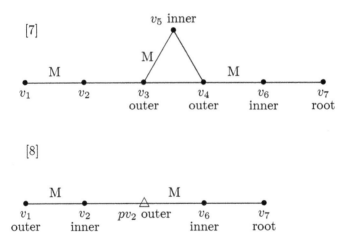

Figure 9.13: In [5] the odd cycle has been collapsed and the algorithm seeks to find an alternating path–and this search is successful in [6]

Figure 9.14: These steps complete the matching because the matching (with root v_7) seen in [7] has a blossom which collapses into [8] and then there can be no augmenting path in [8], because v_7, v_6, pv_2, v_2, v_1 forms a Hungarian tree. [7] is the optimal matching.

9.5.2 Matching in a general graph

The problem of finding the optimal weight perfect matching in a general graph (i.e. one which is not bipartite) is more complex. A description of it occupies seven or eight pages, and draws on linear programming theory which has not been assumed anywhere else in this book; hence a detailed description is omitted. The interested reader will find an elementary presentation in various

	R	S	T	U	V	W
Meg	11	10	8	–	–	–
Neil	–	6	8	6	–	–
Owen	–	–	10	–	11	13
Paul	–	–	–	10	6	11

Table 9.1: Example: The car rental company has six vehicles, and four customers, with profits as shown.

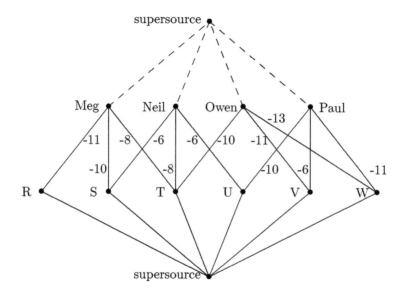

Figure 9.15: The network for the car rental problem. The two vertices labelled "supersource" are the same, but are shown separately for convenience. All the edges are directed "down" the page. The solid edges have lower bound of 0, upper bound of 1. The dashed edges have a forced flow of 1 unit. Unless indicated all edges have zero cost.

texts including Christofides [4]. The method was also devised by Edmonds [6]. It assigns variable weights to the vertices, and to selected sets of vertices. Then, using these values and the weights on the edges, a new, unweighted graph is created. Then the maximum cardinality matching algorithm is applied to the unweighted graph. The process iterates, changing the weights, and consequently the unweighted graph. Eventually, it will converge to a perfect matching of the optimal weight.

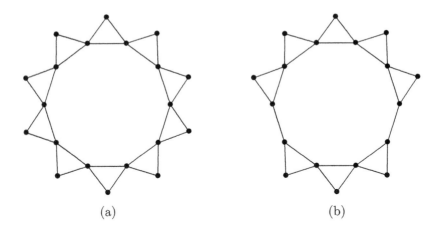

(a)   (b)

Figure 9.16: The star from my Christmas tree, (a) before and (b) after the cat broke it.

9.6 EXERCISES

1. In a certain lecture course, every male student knows exactly k of the female students, and every female student knows exactly k of the male students. Prove that it is possible to arrange all the students in pairs in each of which there is one male and one female.

2. As part of a social event at an international conference, the organizer wants to place as many as possible of the delegates in pairs; within each pair, the delegates must both be able to speak the same language. Show that the organizer's problem can be modelled using a graph, with delegates represented by vertices and edges indicating that the corresponding delegates can speak each other's language, and that the problem is then one of maximum-cardinality matching.

3. The star on the top of my Christmas tree is made of silver wire and looks like Figure 9.16 (a). Find a maximum cardinality matching in this graph. Unfortunately, last year, the cat broke off two of the points, leaving it like Figure 9.16 (b). Show that there is a matching of cardinality 9 in this graph, and demonstrate that such a matching will exist whichever points had been broken off. Hence show that the cat can break off as many pairs of points as she likes, and still leave a graph with a matching with no isolated vertices.

4. Stephen Sondheim's musical *A Little Night Music* is known for the song "Send in the Clowns". It has a cast of ten (four male, six female) and

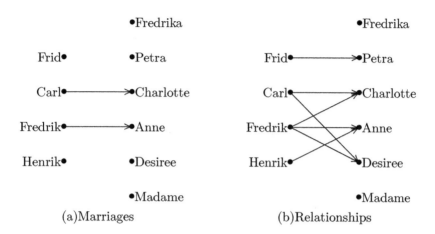

(a)Marriages   (b)Relationships

Figure 9.17: Stephen Sondheim's musical: *A Little Night Music*.

several *Liebeslieder* who are there as singers, and do not take part in the action. There are two married couples, so clearly Hall's (Marriage) theorem applies in Figure9.17(a) with cardinality 2. At the end of the show, several other relationships have been revealed or have developed (Figure 9.17(b)). Show that there is a matching of cardinality 4 in this bipartite graph, and verify that Hall's theorem is true for all sets A of males.

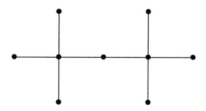

Figure 9.18: Joining two trees

5. A tree with 5 vertices and 4 edges, in which one vertex is the root and all the others are its children, can be represented as a cross with the root vertex in the centre. Two such trees may be joined together to form a tree by identifying one child vertex in the first tree with a child vertex in the second, as seen in Figure 9.18. Joining a further tree to one of the other child vertices gives a tree such as appears in Figure 9.19. Show

Figure 9.19: Joining three trees

that the maximum cardinality matchings of these trees have values 2 and 3 respectively, and prove that if the tree in Figure 9.19 is extended then the maximum cardinality is increased by 1 for each extension.

6. Extend the result of the previous question to show that if M is the maximum cardinality of a planar graph with $|E|$ edges, then the ratio $M/|E|$ can take any rational value r where

$$0 < r \leq \frac{1}{2}$$

What values of $r > \frac{1}{2}$ are possible?

7. Find the matchings of maximum cardinality in each of graphs G_1 and G_2 in Figure 9.20

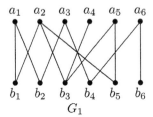
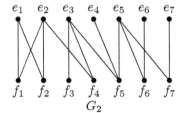

Figure 9.20: G_1 and G_2

8. In Figure 9.21, edges marked "M" are matched. Find an exposed vertex and a matched vertex. Find an augmenting chain that is rooted at vertex v_{17}. Show that one may also find a blossom from v_{17}.

9. What is the maximum cardinality matching possible in Figure 9.21? Is it possible to add one vertex v_{18} and one edge to give a graph whose maximum cardinality matching is 9?

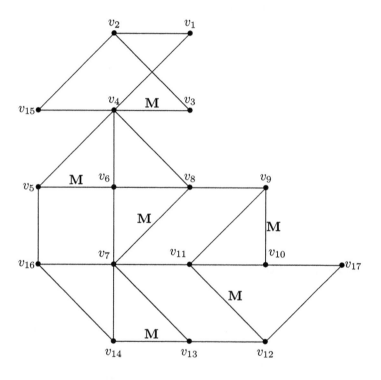

Figure 9.21: A partial matching

10

Postman Problems.

When you have read this chapter, you should be able to

- Describe the family of problems known as "postman problems";

- Know how to solve problems where all edges are undirected, and all edges are directed;

- Know a little about the complexities of other members of the family of problems;

- Recognize the importance of postman problems in commerce.

10.1 INTRODUCTION

Most historians of mathematics reckon that the study of graphs and networks dates from Euler's analysis of a problem known as the "Bridges of Königsberg" in the 1730's. According to the popular story, the people of this European city wanted to know if they could stroll across each of the seven bridges once, and once only, and return to the place where their walk started.

The city, renamed Kaliningrad in the twentieth century, was built on the two banks of the river Pregel, with two islands. The river and its seven bridges looked like Figure 10.1. Léonard (or Leonhard) Euler was one of the most distinguished mathematicians of the eighteenth century and wrote about many aspects of mathematics. He heard about the problem, and rapidly solved it. The people of the city were not planning their strolls to go on a drinking spree but wanted an excuse to show off their fashionable clothes. Euler reduced the problem to a simple diagram, shown in Figure 10.1. This represented the four pieces of solid ground as vertices, and the seven bridges as edges. And then, all that is wanted is a way of selecting a vertex to start from, travelling from one vertex to another using the edges, and returning to the start after using each of the edges once only. This problem is also referred to as a **unicursal**

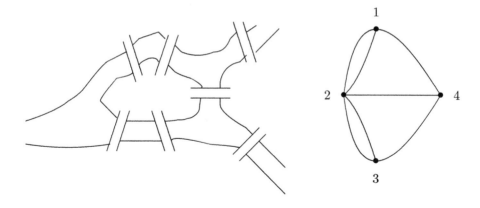

Figure 10.1: The bridges of Königsberg, and Euler's diagram (a multigraph).

problem, where a unicursal network is one in which it is possible to draw all the edges with a continuous line, returning to the starting vertex.

In what seems today to be an elementary piece of analysis, Euler said that each vertex that is passed through must have an even number of edges incident on it, because one edge is used for coming in, and a second edge for going out. The requirement to return to the vertex where the journey commenced means that the start vertex must also have an even number of edges incident on it. Accordingly, a tour of the edges of the type desired is only possible if all the vertices are of even order. A tour like this is known as an **Euler tour**. The graph or network is called **unicursal** or **Eulerian**. By extension, it is possible to select a vertex, travel along all the edges and finish at a different vertex only if the start and finish are vertices of odd order, and all the others are of even order. This is an Euler trail. Since the graph has three vertices of order 3 and one of order 5, there is neither an Euler tour nor an Euler trail around the bridges of the city.

Euler continued his analysis by providing a means for finding an Euler tour or Euler trail, when one exists. The algorithm below yields an Euler tour in a multigraph. It is a recursive algorithm, since step 3 "calls" the algorithm again, for a smaller multigraph.

1. Select a start s and make it the "current" vertex k. All edges "unused".

2. Select any unused edge, incident on k, (k, l) say. Mark it used. Set $k = l$. Repeat until there are no unused edges incident on k, which means that $k = s$. (This is effectively moving along the selected edge, and recording the next vertex reached.)

3. When there are no unused edges incident on s, and there are some unused edges in the multigraph, select a vertex s' which has both used and unused edges incident on it. Construct a postman tour from s' and join it into the incomplete tour.

4. Stop when all the edges are "used"

To find a trail, identify the vertices of odd order, say s and t. Construct an edge (t, s) between them. Start the Euler tour algorithm at vertex t and use the new edge as the first one. This will construct an Euler tour which finishes at t; the Euler trail from s to t is found by deleting the newly created edge (t, s) from the tour.

When there are no Euler tours on a graph, some of the edges must be repeated. One then tries to find a multigraph which has all vertices of even order, with an optimal number of extra edges added to the original. In some cases, there are two costs associated with the edge, the cost of **servicing** the edge and the cost of **traversing** it. Servicing an edge (making the deliveries, collecting the refuse, inspecting the pipes) will usually have a higher cost than simply going from one end to the other, traversing it. However, when there is only one cost, this is normally referred to as the traversal cost for the edge.

This chapter extends Euler's ideas from graphs to networks. It considers the problems where the edges have a weight (cost, distance or time) so that the cost of a tour is the sum of the costs of the edges that are used. A common theme of the methods is that edges may be copied so as to make a graph or network which is unicursal. It also considers problems where the edges are directed. Finally, there are short introductions to areas of current research; nearly three centuries after Euler was born, there are unsolved problems concerned with tours.

10.2   APPLICATIONS AND NOTES

A postal delivery person (usually referred to as a "postman") has to travel along every road (with houses) on his or her rounds. So do milk delivery vehicles, refuse collection vehicles, and inspectors of pavements. Electricity and telephone engineers need to travel along their cables to inspect them. As roads and cable systems can be pictured as networks, there seems to be an obvious way of posing the problem of finding the best way of carrying out the work as a network problem. In most cases of the examples posed here, it is to find the shortest circuit of the network which traverses each of the edges at least once.

A **postman tour** in an undirected network with edges E and edge lengths d_{ij} is a circuit $C = \{(i_0, i_1), (i_1, i_2), \ldots (i_{r-1}, i_r = i_0)\}$ of r edges from E, chosen so that every edge from E occurs at least once in C.

The **length or weight of a postman tour** C, is taken to be $\sum_{j=1}^{r} d_{i_{(j-1)} i_j}$.

The **optimal postman tour** is one whose length is the least possible. Since the only postman tours of especial interest are the optimal ones, it is convenient to drop the word "optimal" in the discussion that follows. The **postman problem** is to find an optimal postman tour, or to show that there is none. The problem is sometimes known as the "Chinese postman problem" because much of the discussion of it follows from a paper by the Chinese mathematician Mei-Ko Kwan in 1962.

Postman tours in directed networks (or mixed ones) are similar, except that the directed edges must be traversed in their correct sense.

For any network, the weight of the optimal postman tour will be at least as great as the sum of all the weights on the edges. Therefore what really matters is to minimize the total weight of repeating edges that must be used (or traversed) more than once. In some cases, edges are used more than twice, and so the definition of the weight of the tour requires that each repetition adds a weight to the total.

Pages on the world-wide-web are hyper-linked to one another. If one is creating a website, then each link should work. So, somehow, every link ought to be checked, by hand or using a computer program. Each page can be represented as a vertex in a directed graph, and each link as a directed edge. If each edge has weight 1, then the test programme which uses the least number of steps is the optimal postman tour on this network. A similar approach can be used to test that every link on a menu, for an electronic device such as a video recorder or mobile phone, or a computer program, is working.

The problems of routing delivery vehicles where there are limits on their capacity includes solving postman problems. Given the high costs of transport, a saving of a few percent on the distance, by finding routes which are optimal will often save hundreds of thousands of pounds (dollars or euros).

In the electronics industry, many circuits have wiring on two layers, because two wires must be separated when they cross. To prevent crossing, a wire on one layer is connected to a wire on the other, by a "via", and the problem of minimising the number of such "via" links is equivalent to a postman problem.

10.3 POSTMAN PROBLEM: UNDIRECTED NETWORKS

For an undirected network, there is always an optimal postman tour.

If all the vertices are of even-order, then there is a solution which traverses each edge exactly once. This is because the graph which underlies the network possesses an Euler tour, and the Euler tour will also be an optimal postman tour. Euler's construction gives a solution.

When there are vertices of odd order, then there will be no Euler tour, and some edges must be repeated. Vertices of order 1 can be removed in a trivial and obvious way. The edges leading to such vertices must be used twice in succession, once going and once coming, and so the edge and vertex can be

ignored in the algorithm. It is therefore assumed that the vertices of odd order
will have degree 3 or more.

Make a list of all such vertices. There will be an even number of them, so
they can be identified as a set $V_{Odd} = \{O_1, O_2, \ldots, O_{2r}\}$ for some $r \geq 1$. The
first stage of finding the optimal postman tour is to find the distances between
every pair in this set, and this can be done with an all-shortest-paths algorithm,
applied to the whole network. This gives the distance matrix and the only parts
of this which are of interest are the rows and columns corresponding to the $2r$
odd vertices; these can be considered as a square matrix of distances with $2r$
rows and columns.

To find the optimal postman tour, some edges must be repeated, and these
edges, together with the original ones, must form a unicursal graph. So there
must be a path from O_1 to another vertex in the set V_{Odd}. Adding the edges
in this path to the original graph makes both vertices of even order. In all,
there must be r pairs of vertices from V_{Odd} linked by shortest paths, making
a graph with an Euler tour. All that is required is the set of pairs whose total
weight is least. This becomes the problem of finding a minimal weight perfect
matching, which has been discussed earlier in the book. However, for small
values of r, it is possible to find the answer by looking at all possible pairings.
When $r = 1$, there is only one pair: O_1 with O_2. When $r = 2$, there are three:

- O_1 with O_2 and O_3 with O_4;

- O_1 with O_3 and O_2 with O_4;

- O_1 with O_4 and O_2 with O_3;

and generally there will be $(2r - 1) \times (2r - 3) \times \ldots \times 3 \times 1$ possible pairings.
One needs to find the best pairing, because the length of the postman tour
will be the total length of all the edges, plus the total length of the repeated
edges, which will follow from one of these pairings.

10.3.1   Postman problem: directed networks

With directed networks, the first concern is that there may not be a postman
tour. This will not happen in reality (one hopes) but in the discussion of
algorithms, one must consider how to stop an iterative process when there
may be errors in the data. If there is a postman tour in a network, then there
will be a path between any pair of vertices, i and j. Conversely, if there is a
pair of vertices for which there is no path, then there cannot be a postman
tour. Accordingly, a simple way of checking whether or not there is a solution
to the postman problem for a given network is to apply an all-shortest-path
algorithm to the network, and find if all the final entries are finite. If so, then
one can proceed.

Vertices in networks with directed edges can be classed as **symmetric** or
not; a symmetric vertex has the same number of directed edges entering as

leaving. A network (or graph) is symmetric if all the vertices are themselves symmetric. Clearly a directed network will not be symmetric if any vertices have odd order.

Symmetric, even order networks present no real problems, since Euler's method can be used to find the postman tour; its length will be the sum of the weights of the edges, as no edge will be repeated, and the only change is that each edge must be traversed in the correct direction.

A postman tour in a network that is not symmetric must repeat some of the edges. To identify these, extend the idea of symmetry by finding the two numbers, d_i^+ and d_i^- for each vertex i. d_i^+ is the number of edges which leave i, and d_i^- is the number of edges which enter i. These numbers must be different for at least two vertices.

Assuming that the edges have length d_{ij}, define the variable $f_{ij} \geq 0$ to be the number of times that the edge (i, j) is repeated. (This is one more than the number of times the edge is traversed.) The objective of finding the optimal postman tour is the same as minimising the objective function $\sum d_{ij} f_{ij}$ while ensuring that the vertices of the graph are visited the correct number of times, represented by an expression

$$d_i^- + \sum_j f_{ji} = d_i^+ + \sum_k f_{ik}$$

This can be rearranged to give

$$\sum_k f_{ik} - \sum_j f_{ji} = d_i^- - d_i^+ = D_i$$

A problem which looks like this appeared earlier in the book. The aim of minimising a sum of the products of a weight (or cost) of an edge, and the number of times that an edge is used by a "flow", subject to specified flow imbalances at the vertices ... is a minimum cost feasible flow problem. So the out-of-kilter method can be used to solve it. The imbalance of flows at the vertices is dealt with by forcing flow into (or out of) them with "input" and "output" edges. Then put infinite upper bounds on the edges of the network, define output edges for those vertices for which $D_i < 0$ to take the flow out of those vertices to some convenient source/sink, and link that same source/sink to the vertices for which $D_i > 0$ with input edges forcing the appropriate flow to their vertices. Then solve the minimum cost feasible flow problem. The values f_{ij} show how many copies of each directed edge need to be added to the original network so as to produce a symmetric network, and then Euler's method can be used to find the optimal postman tour.

10.3.2   Example

Figure 10.2 shows a simple network with directed edges. For this example, all the edges are assumed to have unit length, although the method does not

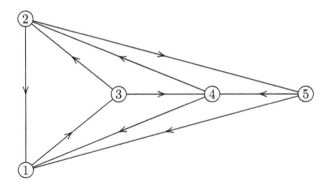

Figure 10.2: The original, directed network; all edges have unit length

assume this. Since there are vertices of odd order, then it is clear that some edges will be repeated. With this assumption, the distance matrix for the network is:

$$\begin{pmatrix} 0 & \infty & 1 & \infty & \infty \\ 1 & 0 & \infty & \infty & 1 \\ \infty & 1 & 0 & 1 & \infty \\ 1 & 1 & \infty & 0 & \infty \\ 1 & \infty & \infty & 1 & 0 \end{pmatrix}$$

and the matrix of shortest distances found by any of the all-shortest-paths algorithms is:

$$\begin{pmatrix} 0 & 2 & 1 & 2 & 3 \\ 1 & 0 & 2 & 2 & 1 \\ 2 & 1 & 0 & 1 & 2 \\ 1 & 1 & 2 & 0 & 2 \\ 1 & 2 & 2 & 1 & 0 \end{pmatrix}$$

so there will be a postman tour.

Calculating the parameters for the five vertices yields Table 10.1. Accordingly, one creates a supersource/sink at vertex 6 in Figure 10.3, and

vertex (i)	d_i^-	d_i^+	D_i	notes
1	3	1	2	
2	2	2	0	symmetric
3	1	2	-1	
4	2	2	0	symmetric
5	1	2	-1	

Table 10.1:

edges $(6,1)$ (flow of 2), $(3,6)$ and $(5,6)$ (flow of 1 each). The direction of these
extra edges may seem wrong. When a vertex has more edges entering than
leaving, there is a new edge created which also enters the vertex. When there
are more edges leaving than entering, the extra edge also leaves the vertex.
Although this may run against intuition, it is correct; knowing that it is the
"wrong" way may even help to remember the direction. The solution to the

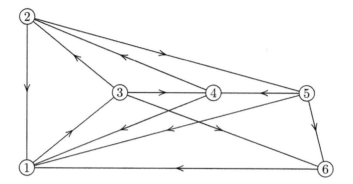

Figure 10.3: The modified network, with a supersource/sink added (vertex 6).

minimum-cost, feasible-flow problem in Figure 10.3 has zero flows all in the
edges except for two units in $(1,3)$ and 1 unit in $(3,2)$ and $(2,5)$. Therefore,
two copies of $(1,3)$ are generated and 1 copy of $(3,2)$ and $(2,5)$ giving the
even, symmetric network in Figure 10.4. Now one can use Euler's method to
find a postman tour.

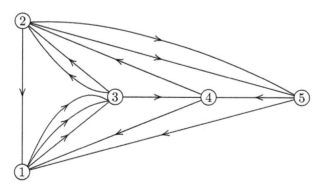

Figure 10.4: The final directed network, with multiple edges added, and the
extra vertex removed.

As this example shows, when the network has directed edges, the optimal

solution to a postman problem may mean that an edge is used more than twice.
For undirected networks, an edge is only used once or twice. There is no limit
to the number of times an edge may be repeated in a directed network. The
network in Figure 10.5 has an edge (t, s))that is repeated N times, for any N.

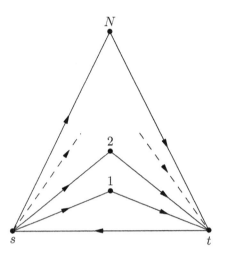

Figure 10.5: In this network, with N pairs of directed edges $(s, i), (i, t)$, the
edge (t, s) must be used N times.

10.4   POSTMAN TOURS IN MIXED NETWORKS.

When the network has a mix of directed and undirected edges, which is probably the most realistic scenario, then analysis can be quite complex. Such
networks fall into 3 categories, depending on the underlying graph:

      case 1) An even, symmetric graph.

      case 2) An even graph which is not symmetric.

      case 3) A graph which is neither even nor symmetric.

    Case 1 is easy to handle because one can separate the directed edges from
the undirected ones, find as many Euler tours as necessary in the directed
edges, as many as are necessary in the undirected edges and then splice the
results together to find an Euler tour in the whole network.
    Case 2 is harder ... and the algorithm follows below.
    Case 3 ... as far as the author knows, there is no efficient solution method
at present.

10.4.1   Postman tours in even, non-symmetric networks

Suppose that $G = (V, E)$ is the underlying graph for an even network, where the vertices are not symmetric. First of all, as in the case of completely directed networks, it is worth checking that a postman tour can exist, by using an all-shortest paths algorithm to see whether there are paths between every pair of vertices. (Nobert and Picard ([16]) proved that for unicursality, even mixed networks must also satisfy the **balanced set condition**. For every subset $V' \subseteq V$, the difference between the number of directed edges from V' to $V \setminus V'$ and the number of directed edges in the opposite direction must be less than or equal to the number of undirected edges between these two sets. In other words, the postman must always have an edge available for getting back to the start.)

Divide the edges into two sets, E_D for the directed edges, and E_U for the undirected ones. Choose an arbitrary direction for each edge in E_U. (It is sensible, but not essential, to make sure that each vertex has at least one edge directed into it and at least one out of it.) This will give a completely directed graph, G_D; calculate the parameter D_i for every vertex. If all the values D_i are zero, then one stops, with a network which is even and symmetric, and Euler's method applies to graph G_D, and hence to the original network with the directions chosen for the edges E_U.

Otherwise, one creates a further network, $G_1 = (\{V \cup \{s\}\}, E_1)$, and solves a minimum-cost feasible flow problem for it.

1. Initialize E_1 with all the directed edges, E_D, with infinite capacity (and zero minimal flow), and cost equal to the edge length for the postman;

2. Add two copies of all the undirected edges, E_U, one in each direction, again with infinite capacity and zero lower bound, and cost equal to the edge length;

3. For each edge in E_U, add an artificial edge, in the direction **opposite** to that chosen earlier; this edge has upper bound 2 and cost 0;

4. Where $D_i > 0$, add an edge (s, i) with a forced flow of D_i;

5. Where $D_i < 0$, add an edge (i, s) with a forced flow of $|D_i|$.

If the problem is infeasible, there is no postman tour. Otherwise, create a further multigraph, $G_2 = (V, E_2)$, using the optimal flow values x_{ij} found in G_1. The edges E_2 are given by the following rules:

- If (i, j) is in E_D, place $x_{ij} + 1$ copies of (i, j) in E_2;

- For the undirected edges, a direction (k, l) was chosen; look at the flow in the corresponding artificial edge (which has direction (l, k)):

  - If it is zero, then the original direction chosen was correct, so place $x_{kl} + 1$ copies of (k, l) in E_2;

  – If it is 2, then the original direction chosen was not correct, so place
  $x_{lk} + 1$ copies of (l, k) in E_2.

The resulting multigraph G_2 will be completely directed, even and symmetric.
Euler's method can be used to find a postman tour on G_2, which is the optimal
tour for the original network G.

Demonstrating that this algorithm works requires proofs that G_2 is even
and symmetric, that the postman tour of G_2 is the optimal tour for G and
that infeasibility of the problem G_1 means that there is no solution for G.

Each of these proofs is reasonably simple. The evenness and symmetry
follows from the fact that all the forced flows in G_1 are multiples of 2, and in
consequence all flows in G_1 will be multiples of 2 as well. The number of copies
of edges created in E_2 will therefore be just enough to correct the imbalance
of edges at each vertex. Flows in artificial edges indicate that the values of D_i
used were wrong, and the algorithm is (effectively) moving the imbalance to
another vertex.

10.4.2   Examples

A worked example of this algorithm requires a larger network than has been
usual in this book. Every vertex in the network must be of even order; if
there are vertices of order 2, then they are not very interesting for the reader,
because the postman tour simply passes through. So any interesting network
must have all its vertices of order 4 or more. This means at least 5 vertices.
As always, the larger the network, the longer the worked example.

Because of this difficulty of illustrating the algorithm, this section has two
worked examples, one a simple network (trivial) with vertices of order 2 and
the second, a larger network with six vertices, all of order 4. Most of the
working will be done pictorially.

Example 1

The first example has two directed edges, and the third which is undirected.
(Figure 10.6.) The answer is obvious, but in order to illustrate the process,
the selected direction of the third edge is deliberately wrong. This is seen
in Figure 10.7, and the calculation shows that $D_1 = -2, D_2 = 0, D_3 = 2$.
Therefore, the network is extended with a supersource/sink, linked by the
edges $(s, 3)$ and $(1, s)$, each with forced flows of 2 units. A reverse edge $(3, 1)$
with infinite capacity, and cost 15, and an artificial edge $(3, 1)$ with capacity 2
and zero cost are added, giving the network G_1 in Figure 10.8.

This minimum-cost feasible-flow problem in this can be solved by inspection, sending 2 units of flow from s to 3, from 3 to 1, and from 1 to s; this
satisfies the constraints, and has zero total cost. Since all the costs are positive on the remaining edges, the total cost cannot be less than zero, so the
optimum has been found.

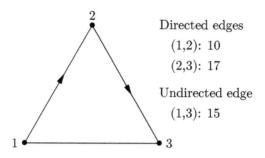

Directed edges

(1,2): 10

(2,3): 17

Undirected edge

(1,3): 15

Figure 10.6: The original, mixed network.

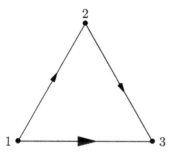

Figure 10.7: Having chosen an arbitrary (and obviously wrong) direction for the edge (1,3).

The interpretation of these flows is that the chosen direction from 1 to 3 was not correct, so that it should be reversed. Since the flows in all the other edges are zero, it is not necessary to make copies of any edges to create an even symmetric graph, G_2, shown in Figure 10.9. An Euler tour can be found in this by the usual algorithm.

Example 2

The original network is shown in Figure 10.10 and the edge lengths are shown alongside. Four edges are undirected, and since vertex 3 is the terminal vertex for three edges, the undirected edge (1, 3) must have direction from 3 to 1. Similarly one forces a direction of from 6 to 4 on another undirected edge; the remaining two edges have been assigned arbitrary directions in Figure 10.11.

In Figure 10.11, the values of D_i are shown for each of the unbalanced vertices. These must be connected to a new vertex, the supersource/sink shown in Figure 10.12. Then solve the minimum-cost feasible-flow problem for

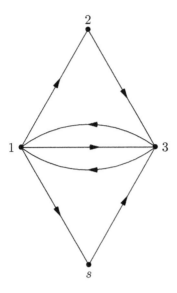

Figure 10.8: Graph G_1 has a supersource/sink s and the extra edges shown.

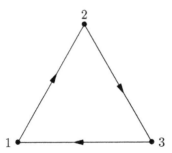

Figure 10.9: Graph G_2, which has the correct direction for all edges, and the appropriate number of copies of each edge (here, one of each).

the network shown there. There are four edges in which flow is forced, and the undirected edges of the original problem have been replaced by three edges each, including the artificial edges defined earlier. The problem in Figure 10.12 has a feasible solution, with flows of 2 units in the edges where flow is forced. It has a flow of 2 units in the artificial edge $(2, 6)$, showing that the direction chosen for this was incorrect, and flows of 2 units in the edges $(1, 5)$ and $(3, 1)$, showing that 2 extra copies of these edges are needed for the multigraph G_2 shown in Figure 10.13. There is zero flow in all the other edges of G_1.

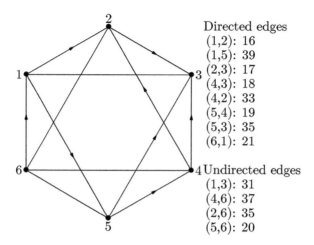

Directed edges
(1,2): 16
(1,5): 39
(2,3): 17
(4,3): 18
(4,2): 33
(5,4): 19
(5,3): 35
(6,1): 21

Undirected edges
(1,3): 31
(4,6): 37
(2,6): 35
(5,6): 20

Figure 10.10: The original network for example 2.

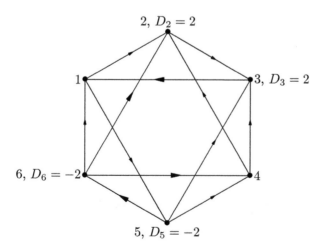

Figure 10.11: The network, with an arbitrary direction chosen for the edges $(3,1), (6,2), (6,4), (5,6)$; these have larger arrows than the others.

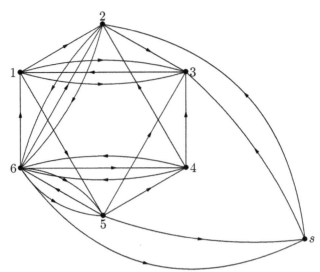

Figure 10.12: The network G_1, showing the extra edges and the artificial edges, with super-source/sink s. The extra edges between the original vertices and s have forced flows of 2 units. The straight edges in the original network have infinite capacity, and cost equal to their weight. One of the paired curved edges has capacity 2 units, and zero cost, the other has infinite capacity and cost equal to its weight. This is the network in which a minimum-cost, feasible-flow problem is solved.

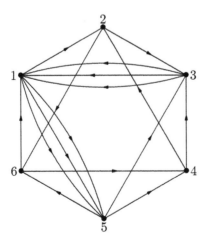

Figure 10.13: The even, symmetric multigraph, G_2 for example 2; an Euler tour can be found by inspection.

10.5 PROBLEMS RELATED TO THE POSTMAN PROBLEM.

There are several problems which are related to the basic postman problem. In this section, there will be brief descriptions of some of these, and comments about them. In general, because the problems are more realistic, then they are harder than the cases that have already been described.

10.5.1 Capacitated postman problem.

Suppose that the parameters on each edge include a length, and a demand; this second parameter may be the weight of mail to be delivered by the postman along the edge, or the volume of rubbish to be collected. The vehicle or postman has a limited capacity, W units. The total demand associated with a walk through the network is the sum of the demands on the edges which are "delivered to" or "collected from". An edge can be traversed without making a delivery, either to allow the postman to reach edges which have not been visited, or because the delivery for the edge has already been made. Here there will often be a distinction between the costs of servicing and traversing the edges. One vertex is the "base".

If the total demand on the network is W or less, then the postman can complete a tour without any difficulty; but if not, then it will be necessary to return to the base, and start again. So, the solution to a capacitated postman problem is a series of walks on the edges of the network, for each of which there is a list of edges to which deliveries are made. The union of these lists is all the edges of the network.

This problem is related to the problem faced by many industries, which have to devise routes for vehicles making deliveries or collections. In such cases, there may be separate trucks or vans, and further complications will be that each one may have a different capacity, and that some edges must be served by particular types of vehicle. (Large vehicles may be barred from using some streets in a congested town, so cannot use the edges of the network that correspond to those streets.)

It can be shown, not only that the problem is NP-hard, but also that the problem of finding a tour that is less than 1.5 times the optimal one is also NP-hard.

10.5.2 The rural postman problem.

The rural postman only needs to traverse a subset of the edges in the network, but other edges are optional and should only be used if necessary. The aim is to route the tour of the network so as to minimize the total distance travelled, while satisfying all the requirements on the journey. The name comes from the problem of delivering mail to separate villages, where every edge must be traversed within the village, but the postman does not need to traverse every edge in the countryside between the villages.

Mathematically, the rural postman problem is based on a graph, $G = (V, E)$ and a subset of edges $E' \subseteq E$. The problem is to find a route which traverses each edge in E' at least once, and has the least total weight. When the graph $G' = (V, E')$ is connected, then Euler's method can be used. If G' is even, then none of the optional edges in $E \setminus E'$ need be used. But if G' has odd-order vertices, then all the edges, including the optional ones, are used to determine the shortest paths between the odd-order vertices. When G' is not connected, more complex algorithms are needed, and heuristic methods are generally used for large problems.

Once again, this is a problem which is based on a practical distribution problem.

10.5.3   The windy postman problem.

The problem where the edge "lengths" are not symmetric is known as the windy postman problem. In this, each edge has two costs, and for some of the edges, the costs are different. The problem is to find a least-cost tour, with the directions of edges selected. It comes from the idea that it is more expensive to go against the wind than with it. Several of the examples of postman problems described are special cases of the windy postman problem.

There are some special cases which can be solved by Euler's method. Otherwise, various heuristic approaches have had reasonable success, based on the linear programming formulation of the problem.

10.5.4   Hierarchical postman problems

The original postman problem, and all the variations that have been introduced, can be further extended by making them **hierarchical**. In addition to all the properties of the edges, one assigns them a priority. Edges with high priority must be serviced before those of lower priority, although it is possible to traverse them in any order. Collection of refuse may be such a problem, since in many cities it is desirable to send the vehicles along the major traffic routes before collecting from quieter, residential streets. In many northern cities in Europe, the USA and Canada, snow-ploughs are sent to deal with the principal roads before those which carry less traffic. In each of these cases, the priority may not always be a strict one. However, mechanical engineers planning routes for cutting tools often have priorities for each cut from which no deviations are permitted. The tool must remove metal in a specified order, otherwise the component will not have the desired properties or shape.

Once again, there are special cases for hierarchical problems which can be solved exactly. In general, heuristic approaches are the best, making use of the linear and integer programmes associated with the network.

10.6 EXERCISES

1. Consider Euler's multigraph based on the bridges of Königsberg. Suppose that each edge has length 1 (1 bridge). Find the all-shortest paths matrix for the four vertices, and hence show that there are two ways of completing a postman tour of length 9.

2. Now suppose that all the bridges cost 1 euro to cross, except for those between vertices 1 and 2, which cost 5 euros. Find the matrix of all-shortest paths, and show that there is only one way to complete the tour at minimal cost.

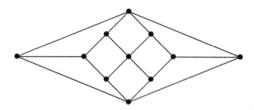

Figure 10.14: The Herschel Graph

3. In the Herschel graph (Figure 10.14), each edge has unit length. Find the optimal postman tour.

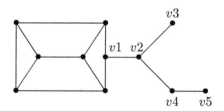

Figure 10.15: Postman network with vertices of degree 1.

4. In many practical examples of postman problems, there are vertices of degree 1, as in the example of Figure 10.15. Show that the problem of finding an optimal postman tour in such a network is equivalent to finding the optimal tour in a network formed by removing a vertex of degree 1 and the edge that is incident on it. Show that this may be done repeatedly. Combining this result with the observation that vertices of

degree 2 are irrelevant to finding postman tours, show that any problem of finding a postman tour is equivalent to the problem of finding a tour on a network (which may be based on a multigraph) where all vertices have degree 3 or more.

5. Which methods of storage of the parameters of a network are suitable for problems involving postman tours, assuming that:

   - all edges are undirected?
   - all edges are directed?
   - the edges are mixed?

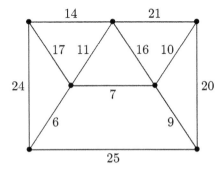

Figure 10.16: Find a postman tour.

6. Given an undirected network with vertices of odd order, it is often the case that the matching can be found by inspection, using the result that no edge need be traversed more than two times. For the network in Figure 10.16, find the shortest paths between all the pairs of vertices of odd order, and show that only one matching is feasible.

11

Travelling Salesperson.

11.1 INTRODUCTION

When you have finished this chapter, you should be able to:

- Recognize a travelling salesperson problem;

- Understand why it is computationally difficult to solve;

- Know about heuristic methods for finding good (but not always optimal) solutions;

- Know about the links with other network problems;

- Know how an exact solution algorithm will work.

11.2 BACKGROUND AND APPLICATIONS

The travelling salesperson problem is one of the best known problems in networks and graphs. This is probably because it is very easy to describe to a non-mathematician, without needing technical language. It is also a problem which has interested specialists from several disciplines; computer science, mathematics, engineering and operational research.

The problem is very simply stated. A travelling salesperson (who is often referred to as being male, but the politically correct name covers persons of either gender) starts in city 1 and must visit customers in each of the cities $\{2, 3, \ldots, n\}$ and return to city 1. Each city must be visited once, and the route taken must be the shortest one that is possible. In the problem one has a matrix (size $n \times n$) of the lengths of a journey between each pair of cities, and the total distance is the sum of the journeys made. It is sometimes assumed that the matrix of distances has been corrected to give the all shortest paths distances.

The main difficulty with the travelling salesperson problem is that with n cities, there are $(n - 1)!$ possible routes, if the matrix is not symmetric, and half this number if it is symmetric. That means that for $n = 10$, with a symmetric matrix, there are just over 180,000 feasible tours, and the number will increase exponentially with n.

Travelling salespersons have often been the butts of jokes, particularly from music-hall comedians of the late nineteenth and early twentieth centuries. Another view can be found in the short stories that Dorothy Sayers wrote about the amateur detective, Montague Egg, who was a fictional travelling representative in the inter-war period. It is a well-established joke that most travelling salespersons do not need to solve the mathematical problem. Nonetheless, there are companies which must distribute goods whose vehicles follow optimal or nearly optimal tours, based on a solution of the travelling salesperson problem (TSP).

Related to the problem of distribution is that of collecting goods. In mechanized warehouses, robots start at a fixed point, collect items from shelves and return to the start. Usually, such warehouses have a series of parallel aisles (like most supermarkets) and the capacity of a robot is limited to a small number of items, so n will be small, and generally it will be sent to the aisles in the order of their layout.

There are production problems in industry which are related to the travelling salesperson problem. Engineering items which need to be drilled are often mounted in a carrier which moves the item relative to a fixed drilling head. Since these movements are repeated for successive items, the carrier must return to its starting position after each complete set of drilling actions. The motion of the carrier can be modelled as a TSP. Another production problem which follows a regular cycle of operations occurs when one machine operates on a repeated sequence of different items, and between each one, the machine must be adjusted or cleaned. The time or cost of these operations depends on the previous item and the next one; for instance, coloured paint, where it may be easier to clean when the transition is from white to pale yellow than when white paint is followed by black paint. Related problems are found in packaging items (it is easier to adjust when the items are similar sizes and are packaged in similar ways than when there are great differences), food production (cleaning after items containing nuts is usually a very thorough procedure), and bottling drinks (strong flavours must be removed before bottling more subtle ones). Preparing hospital operating theatres for a succession of surgery is a further instance. The problem of cutting wallpaper and strips of carpet so that the edges match and the waste is minimal is another variation of the TSP.

11.2.1 Mathematical formulation

Suppose that $G = (V, E)$ is a network, with weight $d_{ij} \geq 0$ on edge (i, j). G is generally assumed to be the complete graph, K_n where $|E| = n$, with weights set to infinity if no edges exist. (In some cases, the weights are calculated by an all-shortest-paths algorithm, so that the matrix of weights has finite entries except on the diagonal. If this is the case, then there may be solutions to the TSP which pass through some vertices more than once, when the shortest path between two vertices passes through a vertex that has already been visited.) Define a **cyclic permutation map** $\pi(i), i = 1, \ldots, n$ to be a map which takes one of the integers $1, \ldots, n$ and generates another, in such a way that the sequence

$$1 \to \pi(1) \to \pi^2(1) \left(= \pi\left(\pi(1)\right)\right) \to \ldots \to \pi^{n-1}(1) \to \pi^n(1) = 1$$

includes all the integers $2, \ldots, n$ once only. Visiting the vertices in this sequence is the **Hamiltonian cycle** defined by the map π. A Hamiltonian cycle is a feasible solution to the TSP; all the constraints are satisfied. With weighted edges, such a cycle produces a **tour** whose weight is the sum

$$\sum_{i=1}^{i=n} d_{i,\pi(i)}$$

(note that this uses the property that each integer occurs once in the sequence, and there is an edge leaving each vertex).

Because each vertex is included in the tour, it does not matter which vertex is considered to be the starting vertex. Hence there are $(n-1)!$ tours, and not $n!$. If the edges are all undirected, then the weight of the reverse tour will be the same, and so the number of tours will be reduced to $\frac{1}{2}(n-1)!$. (But, as was pointed out earlier, this is still a large number.) Travelling salesperson problems are sometimes classified according to whether their weight matrix is symmetric or asymmetric.

Two network problems from earlier in this book have common features with the TSP. These are the minimum spanning tree problem and the assignment problem. As there are simple, efficient algorithms for these, the similarities can be used to help find approximations to the solution of the TSP.

11.2.2   The minimum spanning tree and the TSP

A minimal spanning tree on a weighted graph $G = (V, E)$ has been described as a set T of $n-1$ edges from E, such that the graph (V, T) is connected and the total weight of the edges in T is minimal. Any feasible solution to the TSP will have a set of n edges, TS. Removing any of these will leave a tree. The weight of that tree must be at least as great as the weight of the minimum spanning tree, so the length of the tour for the optimal travelling salesperson tour must be greater than the weight of the minimum spanning tree, and the excess is at least as great as the length of the shortest edge which was not used in the minimum spanning tree. Hence, the algorithms for minimum spanning trees give lower bounds for the optimal solution to the TSP.

But the set of feasible solutions for the minimum spanning tree is much larger than the set of trees that can be formed by dropping an edge from a Hamiltonian cycle. (Cayley's result for the complete graph K_n showed that there are n^{n-2} trees, while there are $(n-1)!$ Hamiltonian cycles, and $n!$ **deletion trees** that can be formed from them.) This is because tree algorithms do not consider one of the constraints associated with the TSP. If a set of edges TS forms a tour, then the degree of each vertex must be 2. So the deletion tree has two vertices of degree 1, and $n-2$ of degree 2. The need to satisfy this constraint is sufficient to make the TSP much more complex than the minimum spanning tree problem, with the consequence that the TSP is NP-complete.

Nonetheless, the idea of finding a minimum spanning tree and adding an edge is attractive as a concept for identifying a lower bound for the solution to the TSP. If one knows that the solution to the TSP is at least some length L, that is often useful information. How large can L be made using minimal spanning trees? One of the most popular ideas is to take one vertex, r, remove it from the network, and then find the minimal spanning tree on the other $n-1$ vertices. Then find the two shortest incident edges that link r to the others, and add these to make a set of n edges, connecting the n vertices. The clever trick is the choice of r. One looks at all n vertices, and makes a list of them, ordered by the sum of the two shortest edges. Now select r to have the greatest value of this rule for ranking the vertices. It isn't a perfect rule, but it is very good in many cases.

For example, taking the seven locations in England's West Country, in Figure 11.1 the ranking is shown in Table 11.1. From this table, Glencot is

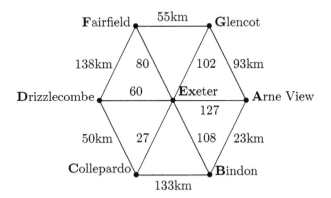

Figure 11.1: The network representing 7 locations in England's West Country, with the road distances between them.

the vertex to be taken as "r". Taking the remaining six vertices, the minimum spanning tree uses edges: $(A, B), (C, E), (C, D), (E, F), (B, E)$ with total

Vertex	edge 1	edge 2	Sum of shortest edges
A	23	93	116
B	23	108	131
C	27	50	77
D	50	60	110
E	27	60	87
F	55	80	135
G	55	93	148

Table 11.1: The lengths of the shortest edges incident on each vertex of Figure 11.1. Take vertex r to be Glencot.

weight 288, giving 436 as the lower bound for the length of the tour of the travelling salesperson. As is evident in Figure 11.2 the seven edges that have been found do not form a tour of the seven vertices. In one of the exercises

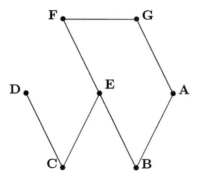

Figure 11.2: The minimal spanning tree on vertices $\{A, B, C, D, E, F\}$, with the shortest edges from G.

at the end of the chapter, one can determine what happens when each of the other vertices is selected as the root vertex r.

11.2.3 Assignment problems and the TSP

The other problem which has many similarities to the TSP is the assignment problem. Suppose that one takes the matrix of distances corresponding to the network of Figure 11.1, and makes the distance from any vertex to itself infinite. This gives a matrix which appears in Table 11.2. The salesperson must leave each of the seven vertices exactly once; he or she must also enter each vertex exactly once. If a matrix of indicator variables x_{ij} has been created

	A	B	C	D	E	F	G
A	∞	23	∞	∞	127	∞	93
B	23	∞	133	∞	108	∞	∞
C	∞	133	∞	50	27	∞	∞
D	∞	∞	50	27	138	∞	∞
E	127	108	27	60	∞	80	102
F	∞	∞	∞	138	80	∞	55
G	93	∞	∞	∞	102	55	∞

Table 11.2: The distances for Figure 11.1

with values 0 or 1, to show that the link i to j is not or is used, then there is a constraint that there must be one entry which is equal to 1 in each row and one in each column. This constraint makes a problem which is very similar to the assignment problem with such a cost matrix. What is the difference between the problems? In the assignment problem, any set of assignments is allowed. So x_{ij} could be 1, and so could x_{ji}. For the TSP, if the salesperson has travelled from i to j, then he or she must visit all the other vertices before returning to i. So there are more constraints on the TSP than simply stating that there is one entry of 1 in each row and column; the combination of entries must define a cyclic permutation.

The need for extra constraints can be seen if the data in Figure 11.2 are treated as data for an assignment. The solution is three "mini-tours", which are not connected, and so are not a Hamiltonian cycle.

The optimal solution using the data in the assignment algorithm is: edges $(A, B), (B, A)$, edges $(C, D), (D, E), (E, C)$, edges $(F, G), (G, F)$ with the tour of the three vertices being taken in either direction. The length of this is 293 kilometres. Because the TSP has more constraints than the assignment problem, this length must be a lower bound to the solution. Once again, the bound has been found using a polynomial-time algorithm; the extra constraints are enough to make the original travelling salesperson problem NP-complete. The three "mini-tours" are shown in Figure 11.3. Later in the chapter, an exact algorithm is described, which takes such a set of mini-tours and finds the solution to the TSP.

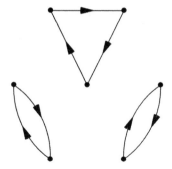

Figure 11.3: Routes found by solving the assignment problem with distance matrix from Table 11.2.

11.3 HEURISTICS FOR THE TRAVELLING SALESPERSON PROBLEM.

Because the TSP is such a complex problem, and because it is fundamental to so many other practical problems, there have been many attempts to find good heuristics for solving it. This section presents a few of these.

11.3.1 Nearest neighbour

Perhaps the simplest heuristic is the "nearest neighbour". One selects a starting vertex, finds the vertex which is adjacent, and whose edge length is least, and which has not yet been visited. This is the nearest neighbour, and one moves along the edge to it, and repeats the process selecting the nearest unvisited neighbour until all the vertices have been visited. Then the final vertex is linked back to the start.

This heuristic is simple, easy to explain, and very poor in its performance. It depends crucially on the selection of the starting vertex. Figure 11.4 shows an example to show that the heuristic can be extremely bad; whichever vertex is chosen as the start, the edge (A,C) must be used, even though its length is many times as great as the lengths of the other edges. In many cases, the heuristic can be improved by selecting each of the vertices in turn as the start, but the diagram demonstrates that this is not always possible. Selecting each of the n vertices as the starting vertex is another (slightly better) heuristic, but it will take n times as long to run as the simple nearest neighbour approach. One can select the best of the tours that are found as the solution to this improved heuristic.

For the seven locations in Figure 11.1, there are three tours found by the heuristic, taking the reverse of a tour to be equivalent to the tour itself. One of the tours is a maverick, because starting at vertex B leads to vertices A,

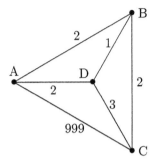

Figure 11.4: A TSP problem where the "nearest neighbour" heuristic is extremely poor. For any starting vertex, the edge (A,C) must be used.

G, F, E, C and D which is not connected to vertex B. The salesperson would need to return to B via E, breaking the constraints of the problem. This is a consequence of Figure 11.1 being an incomplete graph. Of the other tours, the better has length 494 km, starting at any of vertices A, E or F.

11.3.2   A random route.

1. Select one vertex s and make it the start, with all other vertices unvisited;

2. If there are one or more vertices which have not been visited, select one at random and extend the tour to it; if not, discard the incomplete tour and start again;

3. Repeat step 2 until all n vertices have been visited;

4. If it is possible, return to s, and store the tour; if it is not possible, then discard the incomplete tour and start again.

This is a **constructive heuristic**. It can be made into an **improvement heuristic** by repeating it with different random selections.

   With the example of Figure 11.1, the random method will be very inefficient, because there are only 12 tours possible out of 360 possible random selections of the vertices. If this method is to be used to gain useful information, then the salesperson should be allowed to pass through intermediate vertices, using the all-shortest-paths matrix and the assumption that the network has a complete graph underlying it. If this is done, then for example the random route B-F-D-E-G-A-C-B has length 851 km.

11.3.3   The two-optimal method.

1. Create a tour by any method.

2. Select all pairs of edges (i, j) and (k, l) from the current best tour, where the salesperson goes from i to j and from k to l, and $j \neq k$, $l \neq i$. This is done systematically, taking i as the start vertex, and j as the second vertex, with k and l running through all the remaining edges of the tour. Then i becomes the second vertex of the tour, and j the third, and so forth. Calculate whether changing the current tour by going from i to k, taking the tour in reverse from k to j, going from j to l and taking the tour from l to i will reduce the length of the tour or not. If it will, make the corresponding tour the current best one and start step (2) again.

3. Stop when there is no way of improving the current best tour.

This is an **improvement heuristic**. Using the same random tour as the initial solution, the two-optimal method will break it at the links B-F and D-E, create the fresh tour B-D-F-E-G-A-C-B and compare the length of this (868 km) with the initial tour, and reject the fresh tour. Then it will break the initial tour at the links B-F and E-G, creating the second fresh tour B-E-D-F-G-A-C-B, and so on.

The idea of being two-optimal can be extended to three-optimal, at the cost of more computational effort.

11.4   FINDING AN OPTIMAL SOLUTION TO THE TSP.

There are several methods which are designed to find the optimal tour without the general need to exhaustively search all possibilities. Many of the exact methods are efficient, although all of them have the potential to need to examine a very large number of trial solutions.

The method described in this section is based on the assignment problem. As has already been seen, the assignment problem provides a lower bound to the length of the optimal travelling salesperson tour (TST). The assignment problem can produce as succession of mini-tours, each visiting a few vertices in a tour. The idea that is used is that the salesperson must escape from each mini-tour after visiting some of its vertices. So in Figure 11.3, the salesperson follows the mini-tour E-F-G-E. He or she must visit each of these vertices at some stage of the tour; after visiting E, he or she may stay in the mini-tour, or be forced out; if the decision is to stay, then after visiting E and F, the same choice exists. After visiting E, F and G, then the salesperson must break out of the mini-tour. These forced moves can be imposed by changing the matrix of distances, putting weights of ∞ on edges which are forbidden.

The algorithm which follows creates a rooted tree, whose vertices are each assignment problems, with their solutions (the set of edges used, and the total weight). When solved, these assignment problems are interpreted as tours, or sets of mini-tours of the vertices. If the interpretation is a set of mini-tours, then that vertex may be the parent of further offspring vertices. Each of the offspring has the a matrix which has been copied from its parent, with the same forbidden edges, and with the appropriate extra ones, defined below. Banning an edge means that the weight is replaced by an infinite weight, equivalent to imposing a constraint on which edges can be used. Because this means that the children have more constraints than their parent, the weight of the solution to the parent is less than or equal to the weight of the solution to any of the children. So each vertex of the rooted tree gives a lower bound to the weight of its children. This process is known as **branch and bound** and was mentioned as one application of rooted trees in chapter2.

The algorithm maintains a list L_V of vertices which represent mini-tours and have no children, and the weight of the best tour found so far T_B. Then any vertex without children is in one of the categories below:

- A solution representing a tour, with length $\geq T_B$; branching from this is pointless, as the solution to the assignment problem is a feasible solution of the TSP; one of these solutions is the current best tour;

- A solution representing two or more mini-tours, with length $< T_B$; such a vertex is "active" because its children may also have solutions $< T_B$ and may yield feasible solutions to the TSP; all such vertices are in list L_V;

- A solution representing two or more mini-tours, with length $\geq T_B$; any vertex like this can be ignored, because its children cannot give a feasible solution to the TSP that is better than the current best solution;

The algorithm follows:

0: Create a root vertex, whose matrix is the weights of the network, and whose diagonal is filled with infinite entries; set $T_B = \infty$ and list $L_V = \emptyset$; make this the current vertex;

1: Solve the assignment problem for the current vertex, and record the weight of the assignment, W. If the solution represents a tour, do step 2, if not, then do step 3;

2: compare W and T_B and update the best tour on record if $W < T_B$. Then prune all entries in L_V with weights $\geq T_B$; go to step 4;

3: compare W and T_B; if $W \geq T_B$ then this vertex can be ignored; otherwise, create children from the vertex as follows, placing all the children into the list L_V; take N as the matrix for the parent, and suppose one of the mini-tours is $C = \{c_1, c_2, \ldots, c_{i1}, c_1\}$; then it is necessary to examine all the ways that go out of this mini-tour to another one after visiting one of the $i1$ vertices; the following assignment problems satisfy this requirement:

3:1: N with all edges (c_1, y) $y \in C$ deleted;

3:2: N with all edges (c_1, y) $y \notin C$ deleted and all edges (c_2, z) $z \in C$ deleted;

3:3: N with all edges $(c_1, y), (c_2, y)$ $y \notin C$ deleted and all edges (c_3, z) $z \in C$ deleted;

. . .

3:i1: N with all edges $(c_1, y), (c_2, y) \ldots (c_{i1-1}, y)$ $y \notin C$ deleted and all edges (c_{i1}, z) $z \in C$ deleted;

Create these children, and place them in the list L_V;

4: if the list L_V is empty, then stop, with the solution to the TSP equal to the solution to the assignment problem T_B; otherwise, select an entry from L_V and make it the current vertex, and go to step 1.

11.4.1 A small travelling salesperson problem.

to →	1	2	3	4	5	6
from 1	∞	27	11	14	17	35
from 2	31	∞	25	21	26	21
from 3	17	17	∞	35	21	27
from 4	39	35	33	∞	24	13
from 5	23	24	33	21	∞	19
from 6	36	27	40	15	38	∞

Table 11.3: Data for a 6 vertex asymmetric TSP.

Consider the problem with distance matrix in Table 11.3. Solving this as an assignment problem gives $W = 105$ and the assignment $1 \to 3 \to 2 \to 5 \to 1, 4 \to 6 \to 4$, which is two mini-tours. So step 3 is followed, and the mini-tour $C = \{4, 6, 4\}$ chosen. (This choice is arbitrary.) There will be two children, with matrices in Tables 11.4 and 11.5. At the next iteration, the

to →	1	2	3	4	5	6
from 1	∞	27	11	14	17	35
from 2	31	∞	25	21	26	21
from 3	17	17	∞	35	21	27
from 4	39	35	33	∞	24	∞
from 5	23	24	33	21	∞	19
from 6	36	27	40	15	38	∞

Table 11.4: The first child of Figure 11.3 with edge (4,6) banned.

algorithm examines one of these: the data of Table 11.4 gives a tour $1 \to 3 \to$

to →	1	2	3	4	5	6
from 1	∞	27	11	14	17	35
from 2	31	∞	25	21	26	21
from 3	17	17	∞	35	21	27
from 4	∞	∞	∞	∞	∞	13
from 5	23	24	33	21	∞	19
from 6	36	27	40	∞	38	∞

Table 11.5: The second child of Figure 11.3 with edges (4,1), (4,2), (4,3), (4,5) and (6,4) banned.

$2 \to 6 \to 4 \to 5 \to 1$ with $W = 111$, which immediately becomes the new value of T_B. Since the list L_V is not empty, the assignment problem in Table 11.5 is solved, and this has two mini-tours $1 \to 3 \to 1, 2 \to 5 \to 4 \to 6 \to 2$ with weight 115. This vertex of the branch and bound tree can be pruned at once, and the solution found from the first child is the optimal tour, with 111 as the weight.

The method is relatively fast because at each stage one is modifying an out-of-kilter network and solving the problem over again. The slowness of the algorithm comes from two sources. First, the uncertainty of how many circuits there will be in a descendant of a particular vertex; second, the need to explore all vertices which might give rise to an optimal tour, which means that one must check or remove all the vertices that are placed on the list of active vertices.

This solution method is only one of several which work for the travelling salesperson problem . There is a large field of research work on finding methods which are efficient for large problems. What is desirable is to find algorithms which, on average, are both fast and yield answers which are within a small error of the optimum, even if they don't actually find the best solution for every problem that they face.

11.5 EXERCISES

1. Using the data in Figure 11.1 and Table 11.1 find lower bounds for the length of the optimal travelling salesperson tour using the heuruistic based on the minimal spanning tree, selecting each of the vertices $A - - F$ as the root vertex r.

2. Use the nearest neighbour heuristic to find a tour of the nine locations in England's West Country, shown in Table 11.6, using Exeter as the starting point.

3. Using the result of question 2, apply the two-optimal rule to see if there

	Bar	Bri	Dor	Exe	Pen	Ply	Sal	Sou	Tau
Bar	-	34	31	14	37	22	40	47	16
Bri	34	-	20	27	65	42	17	25	16
Dor	31	20	-	18	55	31	13	18	14
Exe	14	27	18	-	37	15	30	38	11
Pen	37	65	55	37	-	26	68	76	48
Ply	22	42	31	15	26	-	44	52	24
Sal	40	17	13	30	68	44	-	8	22
Sou	47	25	18	38	76	52	8	-	30
Tau	16	16	14	11	48	24	22	30	-

Table 11.6: Distances in England's West Country: Bar=Barnstaple, Bri=Bristol, Dor=Dorchester, Exe=Exeter, Pen=Penzance, Ply=Plymouth, Sal=Salisbury, Sou=Southampton, Tau=Taunton.

is any tour improvement which breaks the link Exeter-Taunton. (Do not attempt to do any more than try to use the initial tour found in the question.)

4. Assuming that the weights for a TSP are symmetric, and a vertex s has been selected as the start of the tour, how can the salesperson find the best tour which has the following constraints:

   (a) The tour must visit vertex i immediately before vertex j.

   (b) The tour must visit vertex j between vertices i aand$k$, but these three vertices need not be visited in succession.

   What happens if the weights are not symmetric?

	A	B	C	D	E	F	G
A	-	16	33	43	67	65	95
B	17	-	23	32	52	57	95
C	36	26	-	10	37	34	75
D	49	38	16	-	28	27	73
E	71	56	41	32	-	37	88
F	70	62	39	32	42	-	51
G	97	97	77	75	90	53	-

Table 11.7: An asymmetric salesperson problem.

5. What bounds can be placed on the length of the optimal salesperson tour of the seven locations in Table 11.7?

6. Use a convenient mileage chart from a road atlas, select a small number of locations, and try to find the solution to the TSP for them.

7. There is another class of heuristics which can be used for finding tours; these are the insertion heuristics. The nearest insertion heuristic takes a starting point s, finds its nearest neighbour i, and creates the mini-tour $s - i - s$. Then a third vertex j is found which is the closest to any of the vertices in the mini-tour, and the mini-tour is extended by inserting j to give either $s - i - j - s$ or $s - j - i - s$, whichever is the shorter. The process of selecting the nearest vertex not yet included and inserting it to keep the mini-tour as short as possible continues until all the vertices have been included. Apply this heuristic to the problem in Table 11.6.

8. As an alternative to the heuristic of question 7, the furthest insertion heuristic selects vertices which are the furthest from the mini-tour. Apply this to the problem in Table 11.6. Compare the results of the two insertion heuristics.

12

Tutorial hints

This chapter gives short hints for selected exercises. Exercises which are simply applications of particular algorithms are not included, nor are the "do-it-yourself" networks that have been included in some chapters.

1.1 A search engine given the keys "network" and "diagram" produced over 7000 images when this question was being prepared. The key "multigraph" produces references to the company of that name! Besides the examples mentioned in the chapter, one can find electrical circuits, biological systems and ecosystems, transition diagrams for Markov processes, diagrams of fire alarm systems ....

1.2 Why is this question here? The answer is all about devising an algorithm, and the aim of that algorithm is to eliminate as large a range of potential values of M as possible each time.

1.3 If $M > 1$, the algorithm will need to find the largest and smallest integers; examining each y_i in turn and comparing it with the best found so far will identify these two extremes.

1.4 What will be the limits? Use the definitions, and consider what might happen in the best and worst cases.

1.5 There is plenty of choice of answers here; one can repeat this question for many pairs of vertices, and many of the graphs and networks shown in the book.

2.1 As will be seen in a later chapter, the matrix is one way of storing the weights of edges. Solving this problem using the weights only and no diagram demonstrates that the diagrams are useful for thinking through some of the algorithms.

2.7 (a) Why should this proposed algorithm fail? Which tests in Kruskal's method does it overlook?

2.7 (b) Once again, which test(s) in Kruskal's method does this proposed method overlook?

2.8 If one believes that there is a set which doesn't have this property, then one can either find an example, or prove formally that such a set exists. Alternatively, one can try to construct a tree recursively using the integers $d(v_i)$; find a pair of integers $d(v_j) = 1, d(v_k) > 1$), draw the edge (v_j, v_k) and try to find a tree on the $n-1$ vertices leaving out v_j and with the degree of v_k reduced by 1.

2.9 Taking away edge e_i creates two components to T_1. Why must there be an edge in T_2 which connects these two components?

3.1 The shortest path from A to I via D is made up of the shortest path from A to D followed by the shortest path from D to I. (This simple rule lies at the heart of **dynamic programming**.)

3.3 The time of travel from where I start to Tulsa is equal to the time of travel from Tulsa to wherever I am singing.

3.5 The edge (i, j) with weight d_{ij} will be in one of the shortest paths from s provided $L_i + d_{ij} = L_j$ or $L_j + d_{ij} = L_i$. What are the conditions, if any, which will make $L_i + d_{ij} > L_j$ and $L_j + d_{ij} > L_i$?

3.6 What happens if there is only one edge leaving s?

3.7 Take a very simple example of a graph, such as one with 4 vertices ($\{s, 2, 3, t\}$) and 3 edges ($\{(s, 2), (2, 3), (3, t)\}$).

3.9 This problem is discussed in detail in the Stanford GraphBase [13] which discusses the graphs formed by representing words by vertices, joined by edges if the letters of one word can be changed into the other by a simple operation (such as changing one letter). The book reports that a collection of 5757 five-letter words in English with edges between words which differed by one letter has 671 isolated vertices, but 4493 words form a connected graph with 13619 edges. The diameter of this is 29, the distance between the vertices representing the words *amigo* and *signs*. So there are word ladders between these words, but they will have at least 29 steps.

4.1 There are nine pairs of seats ((L1,R1), (L1,R2), (L1,R3), ...,(L3,R3), in the notation of L=left, R=right), and three couples. Only six pairs are permissible. Hence the problem is represented by the complete bipartite graph $K_{3,6}$.

4.3 The maximum-flow, minimum-cut, theorem means that one can examine all the cuts in this network, and see which depend on X and Y. Then there will be conditions under which one cut will be the minimum-cut.

4.5 One asks: "Were both the ends of edge (k, l) labelled when the algorithm terminated?" If so, then the edge cannot be in the minimum-cut; but what if one end was labelled, and the other not?

5.4 How does one prove that a graph is bipartite? How long is the path between any pair of vertices?

6.3 Suppose that this proposed algorithm is applied to the network in Figure 12.1. What is the shortest path before subtracting c_{min}? And afterwards? What goes wrong?

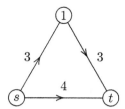

Figure 12.1: Exercise 6.3; what happens?

6.4 Suppose that Shimbel's method were used. After one iteration, the entry in the (i, j)th position will be 0, 1, 2 or ∞. What does an entry of 2 mean? What will the entries be after a further iteration?

6.5 In each case, one asks: "will this change affect any of the shortest paths? If so, why, and what sort of change?"

6.6 What is the longest path that any of the algorithms will find? If there are negative costs, could there be a path which is longer than this?

6.7 If you think that the output will be nonsense, why?

6.8 What stopping rule will be appropriate for a modified form of Shimbel's method?

6.9 Floyd's method depends on checking that all intermediate vertices could be used; vertex 9 will be the intermediate vertex for some shortest paths. . . .

7.3 Look at some of the examples and exercises and (instead of deleting edges) think what happens if an edge is added to the network, once F_{max} has been found.

7.4 Once F_{max} has been found, there will be a residual network. The changed capacity applies to this as well, so the problem is to show that the maximal flow in the changed residual network is a multiple of β.

8.1 (c) This needs to be done iteratively; obviously one of the communications lines in the minimal cut will be the answer, but it may not be the line with largest capcaity.

8.3 Is this a transportation problem?

8.4 Apply the standard questions; what will be suitable vertices?

8.6 To convert a flow problem to a minimal-cost, feasible-flow problem, a return edge is needed. What parameters will minimise the flow in this edge?

8.7 (c) First, see if the optimal solution satisfies the constraint; if so, then no more work is needed; if not, then the problem can be split into two parts.

9.1 Hall's theorem, obviously. Take any subset of m male students; this group has mk edges to female students. Because each of these has k edges to males, there must be at least m female students known to the subset. (The conditions can be relaxed, to be "at least k" instead of "exactly k".

9.4 Are there other plays where such graphs can be drawn?

9.6 There is nothing special about "star graphs" with four edges radiating from a central vertex; there can be any number of edges radiating from such a vertex, and "different" stars may be joined together.

10.4 How does the postman tour visit a vertex of degree 1? And a vertex of degree 2?

11.4 First, test to see whether the constraints matter. Certain edges must be removed from the network to forbid some paths occurring in the optimal tour; how can this be done?

Books and References

[1] Balinski, ML
On a selection problem. Management Science. 1970 **17** p230–231

[2] Boruvka, O.
O jistém problému minimálnim. Práca Moravské P$v$rirodov$v$edecké Spole$v$cnosti 1926. **3** p37–58

[3] Cayley, A
A theorem on trees. Quarterly Journal of Mathematics. 1889. **23** p376–378

[4] Christofides, N
Graph Theory: An Algorithmic Approach. Academic Press. (ISBN 0121743500). 1975.

[5] Dijkstra, EW
A note on two problems in connection with graphs. Numerische Mathematik. 1959. **1** p269–271

[6] Edmonds, J.
Maximum matching and a polytope with 0-1 vertices. Journal of Research of the National Bureau of Standards B. 1965. **69** p125–130.

[7] Edmonds, J
Paths, trees and flowers. Canadian Journal of Mathematics. 1965. **17** p449–467

[8] Edmonds, J and Karp, RM
Theoretical improvements in algorithmic efficiency for network flow problems. Journal of the Association for Computing Machinery. 1972. **19** p248–264.

[9] Floyd, RW.
Algorithm 97 (Shortest Path). Communications of the Association for Computing Machinery. 1962. **5** p345

[10] Ford, LR and Fulkerson, DR
Flows in networks. 1962. Princeton University Press, Princeton, New Jersey, USA.

[11] Hall, P.
On representatives of subsets. Journal of the London Mathematical Society. 1935. **10** p26–30.

[12] Hochbaum, DS and Chen, A
Performance analysis and best implementations of old and new algorithms for the open-pit mining problem. Operations Research. 2000. **48** p894–914

[13] Knuth, DE.
The Stanford GraphBase: a platform for combinatorial computing. ACM Press and Addison-Wesley. (ISBN0-201-54275-7), 1993.

[14] Kruskal, JB.
On the shortest spanning subtree of a graph and the traveling salesman problem. Proceedings of the American Mathematical Society. 1956. **7** p48–50

[15] Land, AH., Stairs, SW. The extension of the cascade algorithm to large graphs. Management Science. 1967. **14** p29–33

[16] Nobert Y., Picard J-C.
An optimal algorithm for the mixed Chinese postman problem Networks. 1996. **27** p95–108

[17] Prim, RC.
Shortest connection networks and some generalizations. Bell System Technical Journal. 1957. **36** p1389–1401

[18] Rawlins, GJE.
Compared to what? An introduction to the analysis of algorithms. Computer Science Press (ISBN 0 7167 8243 X). 1991.

[19] Shimbel, A.
Structure in Communication Nets. Proceeding of the Symposium on Information Networks (1954), (Edited by Jerome Fox), pp199-203. Polytechnic Institute of Brooklyn. 1955

[20] Sollin presented the algorithm during a meeting held in 1961; it was never published in his name, but one of those present reported the algorithm in the book:
Berge, C., Ghouila-Houri, A.
Programming, Games and Transportation Networks. Wiley. 1962.

Index

In general, the first appearance of a term gives a definition or short description.

π-value, 105

adjacency list, 66
adjacent, 4
Ai, viii
algorithm, 3
    alternating tree, 139
    Boruvka, 15
    cascade, 82
    complexity of, 7
    Dijkstra, 73, 139
      choice step, 35, 41
      comparison step, 36, 41
    Edmonds, 137
    flow
      excess-scaling, 100
      FIFO, 101
      highest-label, 101
    Floyd, 82
    Ford, 73
    Ford-Fulkerson, 49, 90
    Hungarian, 119, 142
    Kruskal, 15, 26
    label-setting, 35, 73
    Land, 84
    minimum-cost, feasible-flow, 104, 105, 159, 160
    out-of-kilter, 104, 105, 142
    preflow-push, 94
    Prim, 15
    Shimbel, 81
    Sollin, 16
    tree-growing, 139
    two-optimal, 176
algorithm attributes
    definiteness, 3
    effectiveness, 4
    finiteness, 3
    input, 3
    output, 4
ancestor, 22
arc, 1
assignment, 130

Bach, Johann, 42
balanced-set condition, 159
Balinski, ML, 47, 187
blossom, 140
Boruvka, O, 15, 187
branch and bound, 178

cardinality
    maximum, 130
Cayley, A, 12, 187
chain, 5
    flow-augmenting, 50
Chen, A, 48, 188
Chesterton, G.K., 77
Chinese postman problem, 153
Christofides, N, 145, 187
circuit, 5, 152
complexity
    Dijkstra, 38
    Edmonds, 142
    Edmonds-Karp, 93
    Ford, 75
    Ford-Fulkerson, 53
component, 5

189

Printed and bound by CPI Group (UK) Ltd, Croydon, CR0 4YY

03/10/2024

01040437-0020